P9-DNC-271

Chemicals Control in the European Community

CHEMICALS CONTROL IN THE EUROPEAN COMMUNITY

COMMISSION OF THE EUROPEAN COMMUNITIES
DIRECTORATE-GENERAL XI
ENVIRONMENT, NUCLEAR SAFETY AND CIVIL PROTECTION

Proceedings of a Seminar for Non-governmental Organisations, Brussels, Belgium, 14–15th March 1991.

Publication No. EUR 14385 EN of the Commission
of the European Communities,
Dissemination of Scientific and Technical Knowledge Unit,
Directorate-General Telecommunications, Information
Industries and Innovation,
L - 2920 Luxembourg

The Royal Society of Chemistry Special Publication No. 112

ISBN: 0-85186-197-0

A catalogue record for this book is available from the British Library

Published by:
The Royal Society of Chemistry
Thomas Graham House
Science Park
Milton Road
Cambridge CB4 4WF, United Kingdom

Printed by Bookcraft (Bath) Ltd.

TABLE OF CONTENTS

List of Contributors

Dimitri Angelis, SPC/2, Commission of the European Communities, Consumer Policy Service.

Ingrid Baschab, Division XI/A/3, Commission of the European Communities, Directorate-General for Environment, Nuclear Safety and Civil Protection.

Edward Bennett, Director, DG XI/A, Commission of the European Communities, Directorate-General for Environment, Nuclear Safety and Civil Protection.

Marc Debois, Division XI/A/3, Commission of the European Communities, Directorate-General for Environment, Nuclear Safety and Civil Protection.

Goffredo Del Bino, Head of Division, DG XI/A/3, Commission of the European Communities, Directorate-General for Environment, Nuclear Safety and Civil Protection.

Ronald Haigh, DG V.E2, Commission of the European Communities, Directorate-General for Employment, Industrial Relations and Social Affairs.

Klaus Krisor, Division XI/A/3, Commission of the European Communities, Directorate-General for Environment, Nuclear Safety and Civil Protection.

Georges Mosselmans, DG III/C/3, Directorate-General for Internal Market and Industrial Affairs.

Patrick Murphy, Divison XI/A/3, Commission of the European Communities, Directorate-General for Environment, Nuclear Safety and Civil Protection.

Willem Penning, DG III/C/1, Commission of the European Communities, Directorate-General for Internal Market and Industrial Affairs.

Fernand Sauer, DG III/C/2, Commission of the European Communities, Directorate-General for Internal Market and Industrial Affairs.

Alberik Scharpe, DG VI/B.2/1, Commission of the European Communities, Directorate-General for Agriculture.

George Strongylis, DG XI/B/4, Commission of the European Communities, Directorate-General for Environment, Nuclear Safety and Civil Protection.

Jean Thibeaux, DG V1/B.2/1, Commission of the European Communities, Directorate-General for Agriculture.

Introduction

by Goffredo Del Bino

The aim of the following presentations is to provide a detailed explanation of existing and planned EC legislation in the area of chemicals control. First, however, let us consider two simple questions. Why do we need legislation for the control of chemicals; and more specifically, why do we need legislation at a Community level?

All chemicals are toxic. The difference between them is the dose that is required to cause toxic effects: sugar and salt, if consumed by the kilogram, may be as lethal as milligrams of cyanide or strychnine. In addition to their toxicity, chemicals may also be explosive or flammable or corrosive. Yet others may have deleterious effects on our environment – for example, chlorofluorocarbons.

Given that all chemicals are potentially dangerous, it is reasonable that precautions be taken to assess each chemical and to ensure that appropriate steps are taken to reduce the potential risks associated with its use to an acceptably low level. For instance, many drugs are highly toxic chemicals, but as long as they are packaged correctly, to protect children, and as long as the dose levels are controlled and recommended conditions of use are followed carefully, the risk to the public is acceptably small.

Similarly, many household cleaning products are potentially very dangerous – for example, bleach and ammonia – but provided they are packaged correctly in solid, strong containers with child-proof fastenings and are clearly labelled, with recommendations for safe use clearly displayed, they can safely be used in the home. Nevertheless, we all know that, even with these precautions, many accidents occur every year arising from the use of such products.

One could argue that every manufacturer or importer responsible for placing a chemical product on the market should decide:

- how the product should be assessed;
- how the product should be packaged and labelled;
- how the product should be applied, as well as appropriate protective measures;
- the dose levels and appropriate concentrations for use.

However, even if all manufacturers and importers were to behave responsibly, the above scenario would be a recipe for total chaos: users of products would be presented with an avalanche of information, not knowing whether they could trust any of it. In order to avoid such a scenario, national legislation was developed to introduce procedures and standards concerning

- the authorisation of products such as drugs, food additives, pesticides and cosmetics;
- the classification and labelling of all chemical products;
- the notification of new chemicals;
- the protection of workers using dangerous chemicals.

Such national measures served their purpose until the advent of the European Community. Thereafter, the existence of divergent procedures and standards served as technical barriers to the exchange of chemicals. If, for example, requirements for the packaging and labelling of chemicals differ between one Member State and another, the packaging and labelling would have to be changed every time a chemical crosses a national frontier. This is precisely the type of additional cost that the European Community and the Single European Market were intended to do away with.

Consequently, over the past 30 years, we have developed comprehensive legislation to harmonize procedures and standards in such a way as to allow free circulation of goods while at the same time ensuring a high level of protection for man and the environment. The need to aim for a high level of protection is important and is recognised in the Single European Act. Our objective is not to achieve agreement at the level of the lowest common denominator, but rather to improve the standard of protection across the whole Community.

We therefore need legislation to control chemicals and, in particular, we need legislation at a Community level. We should not, however, forget that the Community has wider international responsibilities. These responsibilities may be rather obvious with regard to the export of dangerous chemicals to developing countries or our role in the protection of the ozone layer and the associated phasing out of fully halogenated chlorofluorocarbons.

Less obvious areas include the transfer of know-how, whereby our legislation is used as a model by countries outside the Community. Such is the case with our new chemicals notification scheme, which has been adopted by many countries throughout the world. The Community is also an active participant in many international fora involved with chemicals control, such as the OECD, the UN Environment Programme (UNEP) WHO, FAO and others.

I should like briefly to explain the development, implementation and control of European legislation, because I am sure that for some participants the functions of the Community institutions and their relationships with the Member States are not entirely clear.

Only the Commission has the power to initiate and propose new legislation at the Community level. Therefore, in the initial stages of developing a proposal for a Council directive or regulation, the Commission departments concerned may consult informally with Member States, individual experts and NGOs concerning the contents of the proposal. When their consultation phase is completed, an official proposal is adopted by the Commission as a whole. This proposal is then communicated simultaneously to the European Parliament and the Council, as well as the Economic and Social Committee. Depending upon the article of the Treaty on which the legislation is based, the Parliament may be consulted once or twice during the consultative procedure.

Eventually – sometimes after a delay of several years – the Council of Ministers will adopt the legislation. Following adoption by the Council, a period ranging from a few months to several years is then allowed for the Member States to introduce the legislation into their national law.

Following the adoption of the legislation, the Commission is responsible for monitoring its implementation in the Member States. The Commission may also be responsible for actively participating in the implementation of the legislation; such is the case for the Community authorisation procedures and notification scheme for new chemicals. Monitoring the implementation of a directive usually involves regular meetings between the Commission and the authorities charged with the implementation of legislation in the Member States.

We have brought together 15 speakers from five directorates-general over 1 1/2 days, to present legislation that has been adopted under Articles 100, 118 and 130 of the Treaty. This is an illustration of the complexity of chemicals regulation and we hope that we will achieve our goal of showing you the ensemble. In spite of the differences in approach we hope it is clear that there is a certain line of logic, a coherence to the system.

On the other hand, we are quite frustrated by the lack of time available to create more opportunities for dialogue such as this one. We want you to provide us with information as well. We need and are looking for more dialogue with NGOs. There will be other seminars, and in particular we want to ask you to tell us what you think we can do to strengthen the dialogue between the Community institutions and the NGOs and to communicate more systematically.

Classification and Labelling of Dangerous Chemicals

by Klaus Krisor

About 25 years ago, governments started to adopt measures to control the use of chemicals. Since then, rules and regulations have been introduced at a world level to protect human beings, among others, against the use of certain chemicals. At a very early stage in the evolution of the European Community, people realised that it would be absolutely essential to harmonize provisions in order to avoid total chaos in terms of trade distribution as well as to avoid technical obstacles to trade which would crop up if individual Member States went their own sweet way in enacting different types of legislation.

A policy on chemicals will obviously have to function at the international level in order to be effective. Differences currently exist among EC Member States, EFTA member states and the other members of the OECD, but EC chemical policy is also generally respected by the six EFTA member states. In other words, 18 of the 24 OECD member states follow more or less the same line in terms of protection of the environment and of humans against the risks linked with the use of chemical substances.

The first steps taken by the Community date back to 1967, when the EC Council adopted Directive 67/548 on the classification, packaging and labelling of dangerous substances. As the years went by, that Directive was amended on various occasions – six times thus far – and it is the so-called 6th Amendment, which the Council approved in 1979, that stands today.

The 7th amendment to this Directive is at present being discussed in Council and is expected to be adopted in the first half of this year. It contains provisions for the notification of new substances and the classification and labelling of dangerous existing as well as new substances.

Marketed substances

The Commission has drawn up a list of all substances which are on the market in the Community. These are the so-called existing substances, defined as those substances which were marketed before 18 September 1981. These form the EINECS Inventory – the European Inventory of Existing Commercial Chemical Substances, which covers approximately 100 000 chemical substances. All non-list chemicals are considered new substances, which must be notified before they can be placed on the market. Notification is explained in a separate presentation; I shall limit myself to describing classification and labelling.

Classification

All dangerous chemicals, both existing and new, must be classified and labelled before being marketed, in order to establish the physico-chemical and toxicological properties of substances which might represent a risk under conditions of normal usage.

Physico-chemical properties are assessed according to the following categories: explosive, oxidising, extremely flammable, highly flammable and flammable. The symbols for these categories are shown in the accompanying figure.

Toxicological properties are broken down into the following categories: very toxic, toxic, harmful, corrosive, irritant, carcinogenic, mutagenic and teratogenic. These categories are defined in the Directive.

Let us look at the current characteristics for classification under the categories very toxic, toxic and harmful. A substance is classified as very toxic if it exceeds a specified LD_{50} (lethal dose for 50% of a test population) for oral absorption in rats or percutaneous absorption in rats or rabbits, or a specified LC_{50} value for inhalation in rats.

A recent development has been the establishment of criteria for classification of a substance as environmentally dangerous. The categories described earlier obviously relate to human health; this category, however, has been created to evaluate the risk these substances represent to various areas of the environment – water, soil and so forth. The 7th amendment suggests a new symbol for that category.

Labelling

Once a substance has been classified under one or more of those categories, labelling follows according to that classification. Just as there are symbols corresponding to these categories (Figures 1 and 2), there are also standardised methods for representing the different characteristics of chemicals, as well as standardised warnings as to the risks involved in their use.

Figure 2 represents what one would typically see on the label of a dangerous substance. The label must contain the name of the substance, the appropriate symbol (representing an indication of danger – "corrosive" in this case), standardised warnings, safety advice phrases and, finally, the name and address of the manufacturer, dealer or importer.

Annexes

The next figure represents a page taken from Annex I of the Directive; it lists those substances subject to unitary classification within the Community. At present, the Annex covers about 100 500 substances, providing the name of the substance, its CAS (Chemical Abstracts Service) number, EC number, chemical formula and the appropriate symbols and standardised risk and safety advice phrases.

The official languages are used in the Annex, with Spanish, German, French, Greek, English, Italian, Danish, Dutch, and Portuguese all listed. It is obviously very important for exporters to have names available to them in the various Community languages, because if they are going to export to different countries they must label exported substances using the name in the language of the country of destination.

Physico-chemical properties

Explosive 

Extremely flammable 

Oxidizing 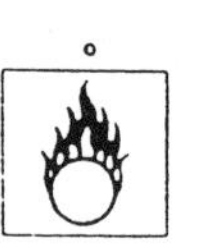

Highly flammable 

Flammable –

Toxicological properties

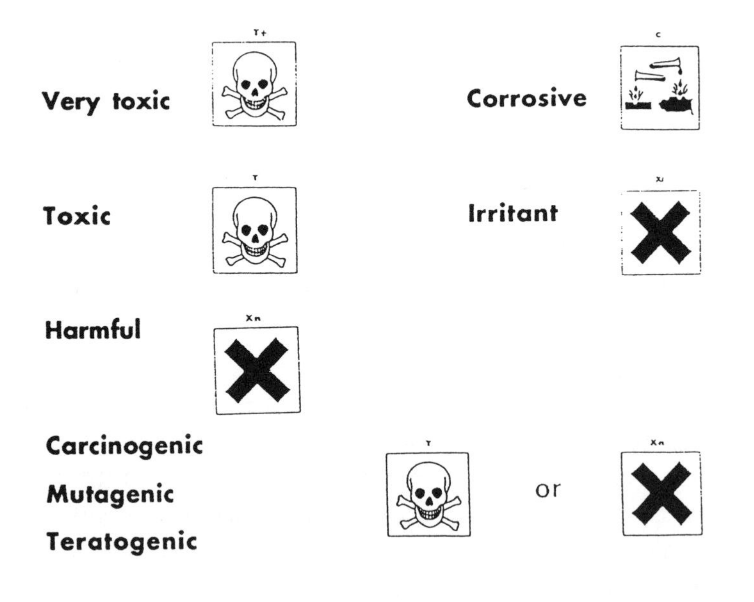

FIGURE 2

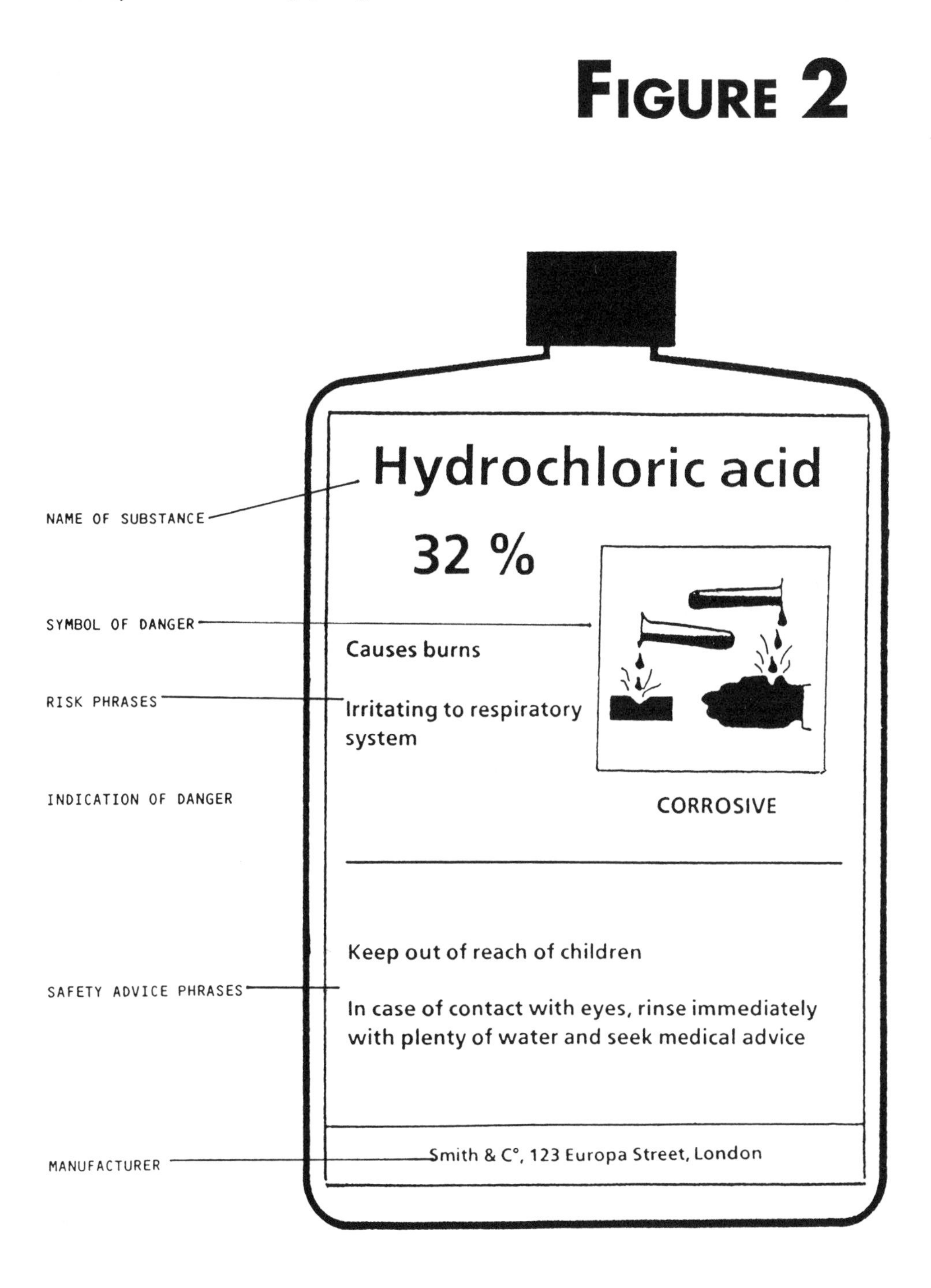
NAME OF SUBSTANCE
SYMBOL OF DANGER
RISK PHRASES
INDICATION OF DANGER
SAFETY ADVICE PHRASES
MANUFACTURER
Hydrochloric acid
32 %
Causes burns
Irritating to respiratory system
CORROSIVE
Keep out of reach of children
In case of contact with eyes, rinse immediately with plenty of water and seek medical advice
Smith & C°, 123 Europa Street, London

Cas No	74-89-5 [1] 124-40-3 [2] 75-50-3 [3]

No 612-001-00-9

Nota C

CH_3NH_2[1]
$(CH_3)_2NH$ [2]
$(CH_3)_3N$ [3]

*ES: Metilamina (mono-(1), di-(2), tri-(3))
DA: Methylamin (mono, di og tri)
DE: Methylamin (mono, di und tri)
EL: Μεθυλαμίνη (μονο-, δι- καί τρι-)
EN: Methylamine (mono, di and tri)
FR: Méthylamine (mono, di et tri)
IT: Metilamina (mono, di e tri)
NL: Methylamine (mono, di en tri)
*PT: Metilamina (mono, di e tri)

R: 13-36/37

S: 16-26-29

The exporter cannot simply translate the name as he sees fit; he must employ the official name as established by Annex I. Exporters must thus look at the EC Directive or the national legislation of the importing country to determine exactly what term is used for a given substance.

The symbols shown in the figure are those used to indicate explosive, oxidising, flammable, highly flammable, extremely flammable, toxic, very toxic, corrosive, harmful and irritant. As mentioned, we have added a symbol indicating environmental hazard in the 7th amendment.

Annexes III and IV contain what are known as "R" and "S" phrases. Annex III lists the R phrases – that is, risk phrases. In our example, for instance, R45 is used. This stands for "May cause cancer", while R46 would mean "May cause genetic damage". These phrases are used for mutagenic and carcinogenic substances. The S phrases (safety advice phrases) are listed in Annex IV. S24, for example, means "Avoid contact with skin", and S25 "Avoid contact with the eyes.

Annex V gives the procedures required for the testing of substances, while Annex VI provides the criteria for the classification of substances under the different categories. Annexes VII and VIII deal in greater detail with the type of tests which must be carried out prior to the classification of new substances.

Amendments

As mentioned earlier, the 6th Amendment is now in force and the 7th is being discussed. These amendments to the Directive cover its legal content – in other words, the articles of the Directive. Thus, if the Council passes a new amending directive, it will amend the legal content of the earlier Directive.

There is also the question of adaptation to technical progress; this involves amendments to certain specific annexes. Amendments to legal content must be given the green light by the Council of Ministers, while adaptations to technical progress are in the hands of the technical adaptation committee and the Commission.

This is the so-called simplified procedure – in other words, when we amend certain annexes we do not need to go to the Council of Ministers. The Commission itself can decide on such amendments once the technical committee has decided what its approach is going to be. So far, there have been 13 adaptations to technical progress.

This is all very confusing – how is one to find one's way? There are six amending Directives and 13 Official Journals, and precisely because it is so confusing, in the 12th amendment to technical progress the Commission drew up a consolidated version which covers all the previous adaptations. Thus, the 12th adaptation replaces all of the previous adaptations to technical progress; that is why the 12th adaptation is voluminous, containing some 2250 pages.

As I mentioned earlier, there are about 100 000 chemical substances in circulation in the Community. According to industry, about 20 000 of those 100 000 existing substances are dangerous according to the lines of the Directive. This means that they could be classified under one of the aforementioned categories. There are approximately 1 500 entries under Annex I, some of which do not actually refer to one specific substance but to preparations containing two or more chemical substances – barium salt, for example, or cadmium preparations – so we can take it that Annex I covers about 5000 specific substances.

I mentioned earlier that there are some 20 000 dangerous substances on the EC market, 5000 of which are contained in Annex I. What about the additional 15 000 which are listed but have not as yet been officially classified? These must be provisionally classified by the manufacturer. In order to give the manufacturer or producer a helping hand with this provisional classification, the Commission has drawn up guidelines which are to be found in Annex VI of the Directive.

Industry wants many more substances classified at the EC level than have been classified so far, since this takes the burden off their shoulders. Consequently, the Commission is at present seriously increasing the number of Annex I substances. We have been working on carcinogenic substances for which there is no unified type of initial classification; that is, different manufacturers have provided various types of provisional classification. We are working on 500 pesticides at present, and on 130 substances which are suspected of causing cancer, as well as a further 100 substances which have been put forward by Member States.

Classification procedure

Very briefly, I should like to deal with how classification comes about at EC level. Let's take an example.

A Member State would like to see a given substance classified. The competent authority in that Member State will gather all the necessary information for classification and will enter it on a specific form. The form is then sent to the Commission, where it is discussed within a working party made up of experts from the Commission, Member States, industry and trade unions; if necessary, scientific experts are also called in by the Commission.

Following these discussions, a proposal for classification and labelling is drafted, which then goes to the Committee on Technical Adaptation, made up of Member State representatives. The Committee will then vote on that proposal. If the vote is positive, the Commission can approve the classification and labelling and the label will then be published in the OJ – that is, a substance is added to Annex I.

Carcinogens

Over the past five years, substances suspected of causing cancer have been given priority. Such substances are broken down into three categories. Without going into details, Category 1 includes substances which are known to be carcinogenic to man; Category 2 covers substances which have caused tumours in animal experiments and which we suspect may cause cancer in man; and Category 3 deals with substances which cause concern for man owing to possible carcinogenic effects, but on which there is not enough available information to give any satisfactory assessment. Information available from animal experiments would not be enough to lead to classification under Category 2.

The actual labelling of carcinogenic substances is carried out on the basis of such an assessment: we use either R45 – "May cause cancer" – or R40, which applies if the substance is classified under Category 3.

CARCINOGENIC SUBSTANCES

CATEGORY 1 Substances known to be carcinogenic to man. There is sufficient evidence to establish a causal association between human exposure to a substance and the development of cancer.

CATEGORY 2 Substances which should be regarded as if they are carcinogenic to man. There is sufficient evidence to provide a strong presumption that human exposure to a substance may result in the development of cancer, generally on the basis of:

- appropriate long-term animal studies
- other relevant information.

CATEGORY 3 Substances which cause concern for man owing to possible carcinogenic effects but in respect of which the available information is not adequate for making a satisfactory assessment. there is some evidence from appropriate animal studies, but this is insufficient to place the substance in Category 2.

CATEGORIES 1 AND 2 Symbol: at least "T"

Phrase: R45 "May cause cancer"

CATEGORY 3 Symbol: "Xn"

Phrase: R40 "Possible risk of irreversible effects"

The Commission also intends to include suspected carcinogenic substances under Annex I; it has therefore called upon Member States to provide information on such substances. In fact, Member States have provided the Commission with information on more than 200 carcinogenic substances. One hundred and eight suspected carcinogenic substances have been listed to date (see figure). The rest are being discussed in working party, and we hope that by the end of this year we will have dealt with all of the substances suspected of being carcinogenic.

Legislative action

Classification and labelling depends on the inherent characteristics of a given substance, independent of its actual use. It does not matter whether it is a pesticide or an industrial substance. We make our assessment on the basis of its physico-chemical and toxicological characteristics – that is, we carry out hazard identification following upon data collection.

Classification, however, is only the first step on a long path towards the detailed risk assessment of a substance (see presentation by P. Murphy). Once that assessment has been completed, we can proceed to legislative action – for example, the banning of a given substance, restrictions on marketing, protective measures in the workplace or measures in terms of dangerous waste. Hazard identification or classification thus leads to an assessment as to whether or not a chemical substance may pose a threat to human health or the environment as a result of its characteristics – that is, risk assessment is a quantitative estimation of the probability of risk.

The final stage in this process is risk management, which is the assessment of alternative regulatory measures and a selection of the appropriate measures to be implemented legislatively.

However, there are cases in which risk assessment is not necessary prior to legislative measures being adopted. In certain cases the situation is so clear that once a substance's characteristics have been established, we can proceed immediately to legislative measures – the adoption of protective measures or restrictions or prohibition. Finger paints for children would be one such example, because these contain substances which have been classified under Category 1 and are carcinogenic. In this case, there is no need for detailed risk assessment.

Conclusions

This is an overview of the activities of the Commission concerning the classification and identification of dangerous substances. Not many aspects of the legislation are not specific to the Community. There are some third countries, in particular the EFTA countries, which have adopted very similar laws. Latin American countries are also developing their own legislation, and some of them have shown considerable interest in Community policy in this area. We for our part are of course greatly interested in getting as many countries as possible to adopt similar if not identical laws, and we consider it incumbent upon us to provide those countries with help.

Of course, we must further develop European policy. At the same time, however, we should encourage and promote legislation in third countries and developing countries and try to coordinate their laws with ours.

Classification and Labelling of Dangerous Preparations

by Georges Mosselmans

Everything that has just been explained in relation to dangerous chemicals is also included in the Directive on the classification and labelling of dangerous preparations. This Directive specifies the same type of labelling, with the same symbols. There are also the same categories of dangers, the same R phrases, the same S phrases. The classification criteria are exactly the same as well – the same biological testing is used as for dangerous substances. I think it is very important to understand from the very outset the more or less exact mirror image of these Directives.

About 100 000 substances are currently available on the market. The number of preparations on the market is approximately 1 million, of which 20–25% are considered dangerous; approximately 1 million preparations are on the market and we can reasonably state that the number of dangerous preparations on the market is also of that order of magnitude.

What is a preparation? It is a mixture of one or several substances. I think this explanation sets the scene, and we can now go into the details of this Directive. I should like to recall that there are two types of dangerous substances: substances which are classified in Annex I, and substances which are classified on a provisional basis by the manufacturer. Both types of substances are covered in the Preparations Directive. Thus, this Directive allows us to cover all dangerous preparations, whether they are already in Annex I or have been provisionally classified by the manufacturer. This was necessary in order to ensure that we were covering everything.

Scope of the Directive

Apart from the pesticides – and there are historical reasons for this – the Directive applies in general to all preparations except those which are already regulated by specific Directives. Some of these Directives concern pharmaceuticals, cosmetics, waste and veterinary and medicinal products. They are described in greater detail in other sections of these proceedings. Furthermore, the Directive does not cover preparations in transit provided they do not undergo any treatment, foodstuffs and animal feeding stuffs or transport.

Classification

The principles according to which preparations are classified are identical to those for substances. Of course, we must also assess preparations in order to identify all their physico-chemical and toxicological properties. Several methods are available to us for the evaluation of health hazards. The first is to carry out toxicological tests. However, it is unthinkable that these tests could be carried out on all preparations placed on the market. We have neither the resources, means, staff, or funds to do this.

Moreover, we must not forget the existence of the Animal Protection Act and the consequent limits to further testing on animals. We therefore have had to think up some generally agreed conventional method. We decided that we would characterise the substance by certain concentration limits, establishing numerical values expressed as a percentage for each of the effects of the substance. There are substances which cause irritation, which are corrosive, which may be lethal, which may cause chronic effects; they may be mutagenic, teratogenic, carcinogenic, etc.

In the Directive, we have listed all of those effects on health, and each of these effects must be assessed. The Directive stipulates that no new tests on animals will be carried out; therefore, we must use the results of tests done in the past. If no such results exist, however, then we use this calculation method based on concentration limits.

These concentration limits are specified in the Substances Directive. For some substances there are personalised concentration limits; when such limits are not given in the Substances Directive, Annex I to the Preparations Directive indicates general limits. In a nutshell, concentration limits are based on two principles: on the one hand, the transformation of LD_{50} values by expressing them as concentration limits so that they have a scientific basis; and secondly, public health considerations, such as the specific effects on health of carcinogenic, teratogenic and mutagenic substances. These limits are purely administrative, designed to provide a maximum level of protection to public health.

Labelling

Having assessed or evaluated all potential dangers of the preparation, we can now label it. It is obvious that labelling a substance is quite easy: one merely indicates the name of the substance. In the case of a preparation, however, one must avoid putting all the names of the dangerous substances on the label. There are a certain number of rules in the Directive which state that all of the substances which are classified toxic, very toxic or corrosive must be indicated when their content is greater than the concentration limit, and it is assumed that a maximum of four to six chemical names are sufficient to identify all of the substances primarily responsible for the major hazards which have given rise to the classification of the preparation.

Thus, labelling involves the same symbols, R phrases, S phrases and names which must be indicated under the Substances Directives. The information about the person who has put the preparation on the market, etc. is also exactly the same.

Safety

There are three other important points which should be emphasised in connection with this Directive. It includes a requirement for safety closures and caps to be fitted to containers containing certain categories of dangerous preparations, for example, preparations labelled as very toxic should be fitted with safety closures; of course also toxic and corrosive preparations; there is an obligation to have a sign which can be detected by the blind; as well as supplementary information over and above the mere label – that is, the safety data sheet.

Together with our colleagues who are dealing with substances and safety in the workplace, we have prepared a Commission Directive which has already been adopted by the Commission and will be published shortly in the Official Journal, in which we set out the information to be provided in the safety data sheet.

The last important point is the article which obliges Member States to set up an "anti-poison centre" – that is, a body which would compile all of the information concerning substances so as to be able to take measures in the event of an accident, as well as to prevent accidents. "Anti-poison" centres have different titles and a different legal position in different countries, so we will simply refer to some body or organisation to be set up by the Member States (some already exist) and which should be provided with all relevant information.

Discussion: Classification and labelling

Q: Are preparations containing several substances considered individually, ignoring synergism?

A: Synergistic effects are not really fully understood, although cumulative effects are. However, work is being done in this area and could result in a future amendment to the Directives.

Q: The use of symbols serves to improve chemical and toxicological recognition, but is anything being done to actually increase knowledge among lay people? And is anything being done to disseminate information about the Community labelling system to university laboratories?

A: Workplace studies are being carried out to see how these symbols are perceived. These are designed to make it possible for the non-scientific public to understand chemical labels, without oversimplifying to such a degree that the phrases used for labelling become meaningless.

However, labels are simply that – labels. It is up to educational authorities to see that the people who use these products are taught to read the labels on them.

Q: Why is the Commission position on organic solvents less stringent than that of Denmark, and on carcinogens in general less stringent that the IARC (International Agency for Research on Cancer)?

A: Some, but not all, organic solvents have been classified as R48, but the Scandinavian countries want all 400 solvents – including substances such as white spirit – to be classified. The EC sees no need for this.

As regards carcinogens, the IARC has dealt with a whole series of substances that the EC cannot deal with, such as alcohol and tobacco, as well as with industrial processes, which are dealt with at the EC level by DG V.

In addition, the EC has chosen to concentrate on substances used in larger quantities.

Q: What exactly is being done to ensure child-resistant closures?

A: Directive 90/35 plans, as of June 1991, to provide for the use of child-resistant stoppers on all highly toxic, toxic and corrosive substances. We are also discussing a second directive dealing with preparations with certain specific characteristics.

Q: Are data sheets to be required for all products? And will long-term effect classification be speeded up?

A: Data sheets had to start somewhere, and the EC has decided to begin with dangerous substances. Full data are missing for long-term effects. It was only in the late 1970s that complex toxicity began to be understood.

This is not a refusal by the EC to analyse long-term effects: the data for doing so simply are not available.

THE COMMUNITY STRATEGY ON ECO PRODUCTS

by Ingrid Baschab

What is an eco-product? All products are produced on the basis of either a natural or a synthetic substance, which means that they automatically will have an impact on the environment. We use raw materials, and we consume energy. Obviously, a technical or manual procedure is required to produce each individual product, and during that stage in a product's life there is also a certain environmental impact. There will be emissions to the air, water and soil. There is also the distribution of products, and the pollution that may result from their transport.

And finally, all products are either used or consumed, which means that they regularly release something into the air, the water and the soil and pollute in a certain way. Thus, sooner or later, all products will, if not completely consumed, become waste. They are not necessarily a risk in themselves when they become a waste product, but on the other hand they may have a major environmental impact because of their volume. All consumer products may pose a risk to the environment, and that risk must be ruled out or limited.

Thus, eco-products, clean products, environmentally friendly products, green products – whatever term one uses – will, according to the Commission proposal, be those products which are environmentally less harmful. That is, their overall impact on the environment will be significantly less than that of other products in the same product group.

The Commission approach

The Commission has selected a comparative approach between competing products which will be used to the same ends and will allow for equivalent use. We are not promoting products which do not raise significant problems for the environment – what we are trying to do is promote products which are less toxic or less risky or less dangerous than alternatives available on the market. As far as we are concerned, an eco–product is one which would help protect the environment as a preventive measure.

I would like to stress the term "preventive". As I noted earlier, this has to apply to all stages of the product's life cycle, from the point of production to distribution, consumption, use and then disposal.

Promotion of the scheme

Promotional measures must aim to get in touch with all the circles involved: with the designers initially, with heads of companies, distributors, trade circles and of course consumers as well – public, private and industrial. We must reach the supplier of the product on the one hand, and the users of the product – the people who create demand – on the other. The latter is particularly important since, in a market economy, information affects competitive position.

The only source of information available to the consumer at present is publicity as such, or labels, symbols, etc. Consumers have become more and more environmentally aware. As a result of that awareness, industry has very quickly started to develop "green" marketing strategies, using publicity and advertising to improve its image and that of its products.

Increasingly, we find products without phosphate, CFCs, mercury, lead, without this or that dangerous substance, or biodegradable. This type of product serves to boost the image of a given company, allowing it to grab a greater market share.

It is therefore clear that the consumer has a major role to play. The consumer can in fact change industry's behaviour, but that sort of change should not be limited to marketing; rather, it should actually change overall product conception. That is why we need an official instrument available to us, to be applied only if rigorous tests have actually been carried out on the product right through its life cycle.

Scope

Some two years ago, the Commission started work on establishing this official instrument – shall we call it the "eco label"? Its use is optional; it is the exact opposite to the negative labelling which is used when a product contains a substance which should be classified as dangerous. But the label on its own will not solve all the problems in this area. This approach is therefore complementary to other measures.

We need certain rules and regulations, provisions, bans, warnings; we have to eliminate fraudulent advertising, etc. We also need to study voluntary agreements. We have some fairly positive examples, such as those relating to mercury in batteries and to CFCs. We must also try to encourage R&D and clean technologies and to organise separate collection for certain products that can be recycled and recovered.

There is one label – the German Blue Angel – which has been in existence for 13 years in the Community. It plays a very important national role and highlights the wider role to be played by consumers. German industry was initially unenthusiastic, but as German consumers became increasingly aware of the environmental impact of certain products and to demand products which had a less serious impact on the environment, industry was forced to react to this public enthusiasm.

In the late 1970s, only 300 products carried this label; now there are over 3000. This example has triggered other schemes, and labels are being used in Japan and Canada. Many other countries are preparing labels as well: these include the Scandinavian countries, New Zealand, Austria and the United States. Within the Community, France launched the NF environment mark about a month ago. The Netherlands is working on a label which may be launched in about a year's time, and Denmark and the UK are already undertaking preparatory work.

Obviously, with the internal market just around the corner, a "go-it-alone" approach is not deemed to be really acceptable. We therefore see a Community approach to this question as preferable.

The proposal

Let us now come to the actual proposal itself. It was submitted to the European Parliament and to the Council quite some time ago. Discussions have started, and some headway has already been made. A year ago, the Council adopted a Resolution which invited the Commission to submit this initiative; and three years ago, in the context of waste management, Parliament asked the Commission to look into the possibility of a label as well as into environmental criteria for products.

This proposal provides an overall framework for the use and management of this. It is in the form of a Regulation; as was noted earlier, a Regulation does not need to be transposed into national legislation but is immediately binding in all Member States on a given date. The eco-label is an economic instrument. We have proposed a regulation because we wished to avoid giving rise to distortions of competition between Member States as a result of the fact that transposition into national legislation takes longer in certain Member States than others.

Objectives

The general aims of this proposal are first and foremost to encourage industry to design and produce products which would create less pollution than other products and, at the same time, to inform consumers and provide guidance for consumer choices by placing certain labels on certain products.

The Regulation also aims to promote rational management of energy and natural resources; minimise emissions into air, soil and water, as well as noise, where applicable; reduce the volume and toxicity of waste; increase product life and encourage the use of so-called clean technology in production, where possible.

An ecological product should not pose a risk to the health or life of the people using it in any way, nor should its characteristics undermine efficient use of the product. In other words, product performance is to be retained.

Product groups

The proposal itself does not contain a list of product groups. Product groups - i.e., products used for the same ends – can gradually be established following the recommendations of trade, industry, consumers and individuals. We must provide access for everyone to be involved in this system. And we are heavily counting on the role of the NGOs, which we hope will be able to suggest areas where there is a need for low-impact environmental products.

Competent national bodies, appointed by Member States, will be informed of such products; they will be the first link in the chain. Our initial aim is to cover consumer products – products that would provide access to the individual in the public or private sector. We have not included services at this stage, but it is possible that the scope of application may be revised after some time has elapsed.

At this stage, we have also ruled out three groups of product: drinks, pharmaceuticals and foodstuffs, all of which are covered by other legislation since they relate to health and safety.

Criteria

We have proposed two types of criteria: general principles, and specific criteria for each individual product. Our general principles relate to respect for essential product requirements and health and environmental safety.

The specific criteria are both qualitative and quantitative in nature. As regards the question of packaging, for example – not a group of products in itself, because we do not want to lead the consumer up the garden path – a particular type of packaging may be labelled green or environmentally friendly or whatever, but the packaging itself may contain a substance which would have significant environmental impact.

Consumers must be sure that a product which bears this label really does represent the best possible alternative on the shelf. Therefore, the criteria to be set up when implementing the scheme have to be very stringent.

These criteria will be drawn up for the entire life cycle of the product. They will, moreover, be applied throughout the Community, because they are being drawn up at the Community level.

The preparatory scientific work will be carried out by the European Environmental Agency, in line with the Commission's proposal. The criteria will be adopted by the Commission, in consultation with an advisory committee. The Commission will in addition draw on the knowledge of in-house and outside experts familiar with the type of product being discussed. Once we have established the criteria and the product groups, these will be published in the Official Journal; preferably, they should also appear in national periodicals, which are more widely distributed and circulated than is the Official Journal.

Awarding the label

When this has been done, companies may apply to use the label. They can do so by submitting an application to the competent national body in the Member State where the product is manufactured or is initially imported into the Community. The body will assess the product to determine whether or not it complies with the criteria. If the results are positive, the body will forward the application to a jury, to be made up of 18 members at Community level. Twelve members will represent the Member States and six will represent interested parties, with one representative each for industry, trade, consumers, environmental groups, media and the trade unions.

Voting will be by two-thirds majority, and an appeal procedure is possible in cases where the Jury rejects an application. The Jury's decision will be published in the Official Journal. The Commission will then check to see that no mistake in procedure has been permitted. If not, the decision becomes official after two months and the national body will draw up a contract with the applicant, awarding the label.

Other provisions

The draft also contains other, more general, provisions. First of all, consumers and industry must be informed as to how the system functions. Member States will organise information campaigns over and above the publication of the Official Journal. There are provisions covering advertising. Any company which has been awarded a label may advertise only that product which obtained the label; they may not use it for other products. Moreover, there should be no possibility of confusion between the Community label and any other existing label.

The logo which has been proposed was the logo of the European Year for the Environment. Of course, like any proposal this will have to be discussed in the Council and final adoption must wait until it has been officially approved by the Council and by the European Parliament.

A review procedure has been established for the scheme. The Commission will look into how the system functions in five years' time. The second point which could be amended or revised is the co-existence of national systems and the Community system. We hope that since the Community label offers the dimension of the Single Market, it will be used and recognised by over 340 million European citizens; the Community label will have the advantage of being used everywhere in the Community, unlike national labelling schemes. This might be particularly relevant for the application field.

Discussion: The Eco-label

Q: Will the label apply to industrial products?

A: Toxicity automatically eliminates a product from meeting Eco-label criteria. Nor will the 80% of chemicals not classified as dangerous therefore become eligible for an Eco-label.

Q: What are the implications of the label for transport?

A: Long-distance transport implications are still being discussed.

Q: Is a link envisaged between the label and the environmental audit being proposed by the EC?

A: The audit applies to non-circulating quantities and is therefore not part of the Eco-label.

Q: What precisely is the interaction between EC and national labelling systems? Will the latter eventually be replaced by the former?

A: The EC does not intend to replace national systems. The expectation is rather that market forces will favour the EC label and eventually lead to the disappearance of national labels.

Q: Environmental groups in the Netherlands are not enthusiastic about the Eco-label. They would prefer a product information scheme, providing consumers with information about all aspects of a product and leaving the choice of product to them.

A: The situation that would result from such a system would be chaos. The Jury will in fact require that all product information be submitted to complete a product dossier, but this information will not then be made public. On the other hand, the Eco-label will include the year of and reason for the award.

Q: What kind of criteria will be used in evaluating manufacturing, transport, distribution and waste disposal?

A: Very specific case-by-case criteria for each product group will be issued when the Council decision on this proposal is final. For now, five pilot studies are being carried out in anticipation of the Council decision, examining environmental impact, energy, use, substances contained in the product, dose and packaging for the product group concerned.

It will also be necessary, however, to determine a cut-off point for consideration of products, which is not necessarily easy. For example, in choosing between oil and palm oil, one has to consider that the former pollutes the air, but the latter will be subject to pesticide applications.

Q: Does the Commission have access to full and correct information about the availability of the clean technology it is encouraging through this proposal?

A: Clean technology is not in fact available for all manufacturing processes. Therefore, the least harmful technology will be sought, together with maximum product life. These factors must be considered together.

Notification of new chemicals under the 6th and 7th Amendments

by Marc Debois

First, I should like briefly to explain the historical background of the notification procedure for new chemicals. I shall then go on to deal with the 6th Amendment, which is the Directive in force at present, the basic idea of notification as such, and other factors which are more or less directly linked with the application of notification. I shall then discuss the 7th amendment, which is currently being discussed in Council. Finally, I shall try to draw a few general conclusions about the notification procedure as applied since 1980 or thereabouts.

Background

Since 1967, when Directive 67/548 was introduced, the Community has had control over chemical substances. The Directive deals with the labelling and packaging of these substances. Notification, or the idea thereof, was introduced under the 6th Amendment, Directive 79/831, which the Council adopted in 1979. That Directive has applied within Member States since 1981, because there was a two-year period left open to Member States in which to implement the Directive through national legislation.

Concepts and aims

Basically, the purpose of notification is to gather minimum data on each substance on the European market, including the identity of the substance, toxicological data and the physico-chemical characteristics of the substances. This information also assists in other Community procedures such as locating the substances discussed at length in earlier texts. In certain cases, it also makes it possible to adopt more restrictive measures regarding the use of certain substances.

I shall also touch upon some of the definitions given in the Directive, as I think this will help provide an overall picture of the 6th Amendment. We tend to think in terms of marketing, and in fact the Directive does concern products being marketed within the EC. It does not apply to non-marketed substances, substances in transit or exports to third countries. On the other hand, the Directive does cover substances imported into the Community. The transport of dangerous substances remains outside the scope of the Directive because it is covered by other Community legislation.

Let us consider the substances themselves. We are talking about pure substances; this means that under the notification procedure all preparations on the market are excluded. But if a new substance is imported into the Community contained in a preparation, that substance must be notified. Thus, there is not notification of the preparation as such, but rather of the new substance which makes up a certain percentage of the preparation.

The Directive applies only to new substances. Existing substances – that is, those which were already on the market before 1981 and have therefore been included in the EINECS list of existing substances – will be excluded from this notification procedure. The notification thus refers only to new substances placed on the European market – in other words, substances not on the EINECS list.

No list of substances covered by notification has been published since the Directive entered into force. In 1991, we will for the first time have official publication of a list of those substances which have been notified since 1981. The data provided for each substance will include its EC number, notification number, trade name, chemical name (when confidentiality is not demanded by the notifier) and, for dangerous substances, official classification in Annex I. This list, which will be updated annually by the Commission services, will supplement the EINECS list.

Notification

In principle, new substances will be notified if they are marketed in quantities greater than one tonne per manufacturer per year. There are, however, a few exceptions to this rule; these include medicinal products, narcotics and radioactive substances, which are covered by other directives. Other substances excluded from the procedure include polymers and similar substances – for example, those containing less than 2% of a new monomer; substances which are being researched and analysed; and substances being used for R&D purposes, for which there is a waiver under certain circumstances.

The notification procedure is laid out in the accompanying diagram. In the case of a European producer, that producer must notify. If the producer is not located within Community territory, then the person who imports the substance will notify. There may be more than one importer in the Community; in that case, each importer is required to notify the competent authority in the Member State where the product is placed on the market for the first time. Thus, each individual responsible for notification provides the competent authority in the appropriate Member State with a full dossier, covering items such as test results, location data, etc.

In the case of an imported substance, the second person responsible for notification may provide what is called a "multiple dossier" to the competent authority. That is, they provide identification of the subject but they may make reference to the dossier supplied by the first importer for the eco-toxicological studies, on the condition that an agreement has been reached between the two parties.

Let us now look more closely at the role of the competent authorities. The notifier submits the dossier to the competent authority, which must then check that the dossier meets the requirements of the Directive. It checks whether the information is complete, if tests have been carried out correctly and whether all the provisions in the Directive have been respected. The competent authority will then draw up a summary of the dossier. This is a standardised dossier that goes to the Commission, which once again checks it and immediately provides all other Member

States with that summary. The information compiled is thus available to the competent authorities in all Member States.

Questions may well be sent in from other Member States, so there is a shuttling back and forth of questions and requests for further information. This is known as the "follow-up" to a dossier. All of this information and these questions are channelled through the Commission, which is responsible for coordinating the various competent authorities.

The information provided in a follow-up may not in all cases solve all the problems posed or answer all the questions raised – for example, points of interpretation may remain unanswered. Any problems not solved under this follow-up procedure will go on to the Directive's management committee, which brings together the Commission and Member State representatives. It is the management committee which will provide a joint answer to those questions which the follow-up procedure has not been able to solve. There are also sub-groups set up to deal with more specific questions on items such as testing methods, risk assessment, categories of use and so on.

Let us look at the repercussions of notification. Forty-five days after the substance has been notified to the competent authority, the person responsible for notification is authorised to introduce the substance to the European Market. Notification is thus basically a type of passport which allows a product freely to circulate on the Community market.

Now let us see exactly what this notification covers. There are three different levels, which really depend on the amount being marketed. The first part covers mini-notification; I shall deal with this further on, as it is not part of the notification established by the 6th Amendment.

Notification levels

The first basic level – or level "zero" – covers those substances which are marketed in quantities of one tonne or more per year per producer. In this case, the producer (or importer) provides a technical dossier, which is described in detail in Annex VII. It covers the identity of the manufacturer and the notifier, the identity of the substance, CAS (Chemical Abstract Service) number, chemical composition, impurities, scope of use and production levels planned for that year and for the future, as well as precautions for use and the possibility of rendering the substance harmless.

Information must also be provided on the physico-chemical properties of the substance – flammability, flash point, etc.; the results of toxicological studies; acute toxicity; mutagenicity and carcinogenicity; eco-toxicological studies, including acute toxicity for daphnia and fish; and the bio-degradability of the product. There should also be a proposal for classification and labelling of the substance, applying the sort of criteria described in earlier texts.

The choice of information demanded is largely based on the procedure established by the OECD for the assessment of marketed products – as noted, these include identification information and general information on the substance, but the technical information on test results may be waived if these results are already included in a previous file.

Level 1 covers substances marketing in quantities of more than 100 tonnes per year per notifier, or 500 tonnes per notifier in total. In this case, additional toxicological or ecotoxicological studies would be required, as established by Annex VIII of the Directive – these include fertility studies, additional mutagenesis studies and toxicity studies, as well as tests for algal toxicity, prolonged toxicity for fish and daphnia, and species accumulation.

There is an intermediate level established by the Directive. This is when the quantities of a substance marketed reach 10 tonnes per year per notifier or 50 tonnes per notifier in total. In this case, the Directive states that a programme of complementary studies should be drawn up by the national authorities in cooperation with all the other authorities (Annex VIII of the Directive).

Level 2 concerns substances marketed in quantities of over 1000 tonnes annually per notifier or a cumulative amount of over 5000 tonnes per notifier. In addition to basic toxicological studies, supplementary toxicity and eco-toxicology tests are required. The data requirements for level 2 are also described in Annex VIII of the Directive.

Test methods

I shall briefly describe the test methods required for notification. The tests carried out on substances must use the standardised methods established in Annex V of the Directive. A thorough review of Annex V has been carried out in recent years and will be published in the first part of 1991. The purpose of the review was to adapt the Annex to technical progress and to introduce a whole series of amendments aimed at reducing the number of laboratory animals used in these tests, in compliance with the 1986 Directive on the Protection of Laboratory Animals.

In particular, the new Annex V stipulates a fixed-dose annex to replace the LD_{50} method traditionally used in toxicity tests. It should also be noted that tests carried out on a substance must comply with the 1986 Directive on good laboratory practice.

Finally, I shall briefly describe the system of mini-notifications. Notification is still required when a substance is marketed in quantities of less than 1 tonne per year per manufacturer, but at this volume the Directive specifies that notification will remain at the national level. Each Member State will be informed of substances that have not been subject to harmonisation. There is also a minimal exchange of information as regards, for example, trade name, proposed classification, etc.

The accompanying Figure 1 shows the number of notifications received since 1983 – the initial figure, in 1983 was 19; this has risen to 350 in 1990. There is thus a clear upward trend. The next Figure 2 gives the accumulated number of notification files received. This relates to substances marketed in quantities of more than 1 tonne per year per manufacturer – basic notification, in other words.

These statistics are linked to the number of notification dossiers. This is an important point since it means that if a substance is imported, there may be more than one file for that substance. At the end of 1990, taking account of this duplication, we had received notification of 535 substances marketed in quantities of more than 1 tonne.

The following Figure 3 shows comparative changes between first notifications and repeat notifications since 1983. Clearly, the ratio

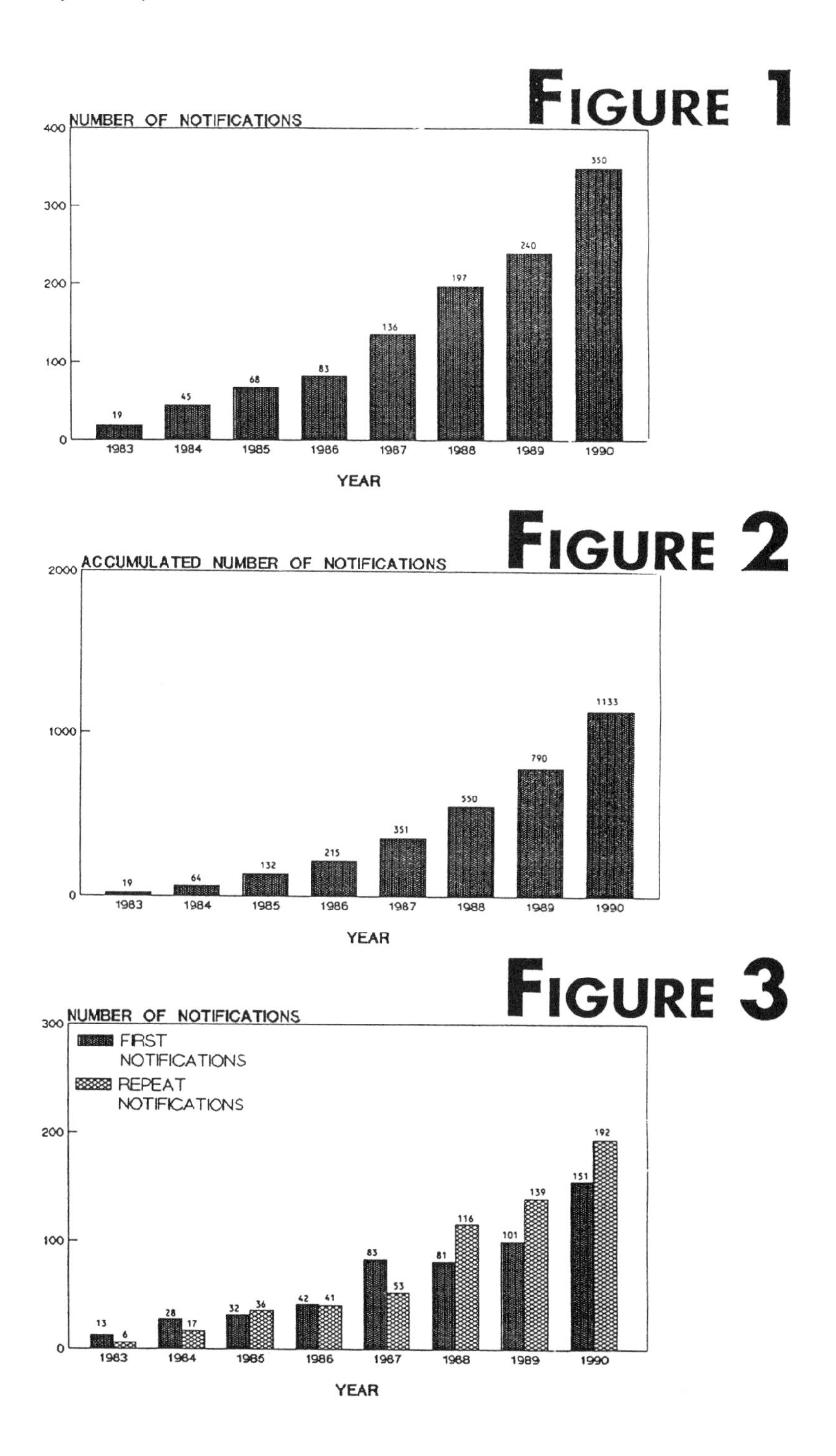
FIGURE 1
NUMBER OF NOTIFICATIONS
400
300
200
100
0
19
45
68
83
136
197
240
350
1983
1984
1985
1986
1987
1988
1989
1990
YEAR
FIGURE 2
ACCUMULATED NUMBER OF NOTIFICATIONS
2000
1000
0
19
64
132
215
351
550
790
1133
1983
1984
1985
1986
1987
1988
1989
1990
YEAR
FIGURE 3
NUMBER OF NOTIFICATIONS
FIRST NOTIFICATIONS
REPEAT NOTIFICATIONS
300
200
100
0
13
6
28
17
32
36
42
41
83
53
81
116
101
139
151
192
1983
1984
1985
1986
1987
1988
1989
1990
YEAR

has switched. Until 1987, we received predominantly first notifications; now, on the other hand, we regularly receive multiple notifications. As might be imagined, this can lead to certain problems.

The Figure 4 shows the evolution of this development; it is quite clear that the number of repeat notifications is increasing as compared to the number of first notifications.

Mini-notifications now take the form of reduced notification – that is, notification at national level for amounts of less than 1 tonne. The number of notifications of low-volume substances has shown a parallel development to the number of files received – there has been a major increase. The accumulated number of notifications by the end of 1990 was 3129.

Classification and labelling of new substances is carried out on the basis of a classification proposed by the notifier; broken down according to risk category, this classification is reviewed and sometimes verified by the competent authority. The accompanying Figure 5 applies only to those categories of substances that are dangerous to man as the result of their physico-chemical characteristics. We have no figures as yet on the threat from substances that represent a risk to the environment.

The total number of substances taken into account for this graph is 449. If each number at the top of the column is then divided by 449, one obtains the percentage figures: 59% for no label, which means no proposed classification. Working downwards in decreasing order are the categories irritant, sensitising, harmful, corrosive, toxic, etc. The highly toxic and toxic substances are not in the first part of this graph, which is quite reassuring, since it means that the majority of substances would not be classified.

Modifications to the 6th Amendment

The 6th Amendment has now been in force for some 10 years. We have seen quite a major increase in the number of dossiers and of substances notified. I think it is therefore fair to say that the system has worked well and that we now have a minimum amount of data available for every substance that has been placed on the market. We are thus able to carry out the basic assessment needed for the classification and labelling of substances.

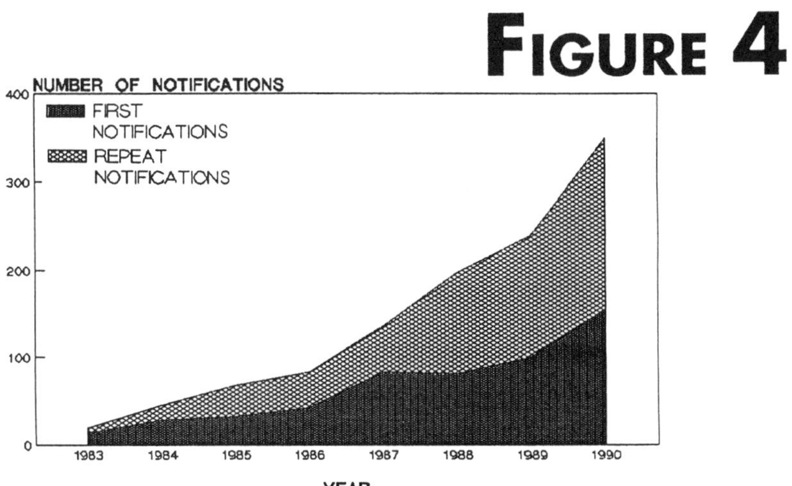

FIGURE 4

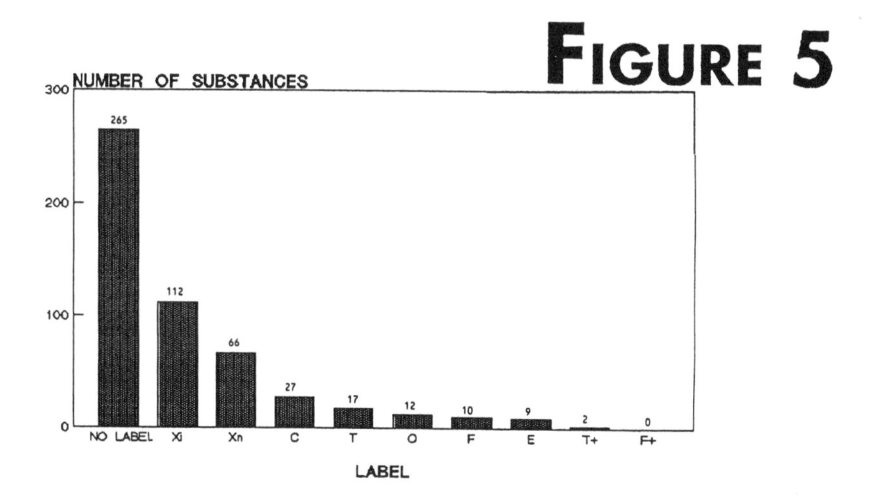

FIGURE 5

More general problems now dealing with those within the competent authorities has meant that it has been able to take a more harmonised approach to notification within the Member States. The 6th Amendment aimed to ensure a unified system of rules applicable to dangerous substances throughout the Community. As demonstrated above, the Directive has been quite successful. However, the 6th Amendment contains certain anomalies and loopholes as well; to correct these we felt there was the need for a 7th amendment.

The Commission tabled the 7th amendment in 1990. It has already been discussed in Parliament and is now up for discussion in Council. The 7th amendment is not a review of existing legislation; rather, it is an attempt to rule out the sort of anomalies which we noticed when we implemented the 6th Amendment.

The first major change under the 7th amendment is that we are now suggesting harmonisation of the notification procedure for low-volume substances (less than 1 tonne per year per manufacturer). As I mentioned earlier, there were problems linked to such notifications because they were carried out at the national level.

Such differences clearly run counter to the whole idea of the Single Market. The suggestion was therefore made that we should try to harmonise these limited notifications under the 7th amendment. However, we will not require the same amount of data for these limited notifications. Instead, there will be two levels of low-volume substances (from 100 kilos to 1 tonne and from 10–100 kilos) subject to a reduced notification procedure.

The second main change deals with extra-Community producers and the problems associated with them. A major loophole of the 6th Amendment is that such producers may use various importers for the same substance. This gives rise to two problems. In certain cases, we have received over 30 notification dossiers for a single substance; these multiple notifications have posed administrative problems.

At the same time, protecting human health and the environment becomes more difficult, because it is possible for a manufacturer to avoid high-volume testing requirements. If a Community producer markets 100 tonnes/year of a given substance, he will be required to provide Level 1 tests. But a non-Community producer can market a substance through 10, 20, 30 – however many he needs – importers in batches of less than 100 tonnes per year and thereby be exempt from Level 1 testing.

This represents discrimination against Community manufacturers, as well as being unacceptable in terms of environmental and human protection, since these products would have a large market presence without having been duly assessed. In order to correct that problem, the 7th amendment proposes that each non-Community producer nominate one person to the Community responsible for notification of the aggregate import.

In addition, a new symbol for risk to the environment will be introduced under the 7th amendment.

One final, important objective is the introduction of risk assessment of the new substances. Data collection was the primary aim of the 6th Amendment. Of course, classification also came under that amendment, but this classification was part of a process of collecting information for an initial assessment of substances. However, there was no link between that process of data collection and the adoption of restrictive measures affecting dangerous substances, because such measures come under Directive 76/769 relating to restrictions on the marketing and use of dangerous substances. There are some further modifications of less importance as well.

In conclusion, the 6th Amendment was efficient in collecting data to make possible an initial identification of problems. The 7th amendment will go further, tackling certain problems of implementation by introducing a new understanding of risk assessment.

Discussion: Notification of new chemicals

Q: What will be done about the huge number of unmarketed new substances that are used in research, etc.? Will data on these be made public?

A: Since 1981, about 400 full base set notifications of new chemicals have been made; in the same period, there have been some 3000 limited announcements (< 1 tonne/year/notifier), some of which were multiple notifications.

In that same period, the Chemical Abstracts registry listed some 5 million chemicals. In the US, many more of this number have been notified, because the US scheme (TOSCA) calls for pre-marketing notification. The EC, however, has decided to deal only with marketed substances, since chemicals not placed on the market are not seen as affecting the protection of man and the environment.

As regards public disclosure of information, the EC has decided to allow manufacturers some measure of confidentiality. Moreover, notification dossiers are huge and do not allow for reasonable dissemination of all data. Nonetheless, all data applying to health and environmental protection are made available to the public.

Q: Industry's desire for confidentiality could be seen as over-riding environmental and consumer requests for information. There is, after all, no legal reason for toxicological information to remain confidential. Indeed, periodic reviews of full information by university experts would be useful; information as currently given is only an abstract of test results.

A: Industry is unwilling to spend 5–10 years and huge sums of money on a product if all the data will immediately be given to competitors. A balance between interest groups must therefore be struck, and the EC feels it has achieved that balance.

Q: Are mini-notifications sent to the Commission?

A: No. They are sent only to the Member States, which circulate them among themselves. However, since tonnages have been lowered under the 7th amendment, these will now become notifications.

THE SYSTEMATIC EVALUATION OF EXISTING CHEMICALS

by Patrick Murphy

Preceding presentations have already described the Community's new chemicals notification scheme and have noted that between 500 and 600 new chemical substances have been notified over the past 11 years. Beyond these, however, there are over 100 000 existing substances, which gives some idea of the relative scale of the problem confronting us as regards that category of substances.

We have a procedure for new chemicals whereby we receive information about them before they are placed on the market. Thus, we know what the properties of a chemical are before it is marketed. This is not the case for the over 100 000 chemicals that were placed on the market before 18 September 1981 – the critical date when the notification scheme came into force.

Existing controls

These existing chemicals are not completely ignored; we do in fact deal with problems of classification and labelling. As Dr Krisor has noted, there are some 2500 substances listed in Annex I to the Directive – that is, 2500 substances which are classified and labelled. There are also some substances which are controlled in specific contexts – for example, in the context of the workplace.

There are also substances which are controlled under another Community directive, which is concerned with restrictions on marketing and use. This is a very short list, and no substance on that list would come as a big surprise to anybody interested in chemical control. Basically, we have shut the door after the horse has bolted, for this list includes PCV and asbestos, cadmium and some lead and tin compounds which are bad actors in anybody's book.

Now, after they have had their effects, we have all decided, in the light of experience and the proof of our own eyes, that they have been damaging the environment and that, yes, we must shut the door and ban them. This is known as the a posteriori approach to chemicals control. We must now move towards a new era of chemicals control.

The accompanying figure shows the list of existing substances. There are 100 000 substances (100 116, to be very precise) in this inventory, which has been published in all nine Community languages. We use this list as our base for identifying what substances should be referred to as existing chemicals. The substances in this list are thus officially the substances which are described as existing chemicals in the Community.

In terms of action to ban or restrict the use of these chemicals, shall we wait until one of them turns out to be another bad actor and is found to be causing problems of environmental contamination or adverse health effects? This rather ad hoc, after-the-event, approach to chemicals regulation has one advantage in that everyone can usually agree to ban such substances, but normally it is already too late if everybody can agree to a ban.

The fact that there are 100 000 of these substances causes certain problems in that we cannot deal with them all at once. Neither industry, the regulatory bodies, the testing laboratories nor the scientific fraternity can deal with 100 000 substances all at once. We have to find a way to break these down into smaller units.

Proposed categories

The Commission has recently made a proposal for a Council regulation on the assessment and control of the environmental risk of existing chemicals. This regulation, which was adopted by the Commission late last year and is currently being discussed in the Council, breaks existing chemicals down into a number of distinct groups which we are proposing to tackle on a slice-by-slice or bite-size basis.

First of all, we have an annex to the regulation relating to all chemicals which we know about that are produced in quantities of

"European Inventory of Existing Commercial Chemical Substances"

contains

100 116 Chemical Substances

marketed for the first time before

the 18th September 1981

EINECS includes

- The EINECS number
- The name of the chemical
- The molecular formula
- The CAS number

above 1000 tonnes per annum. For those chemicals, we would like to receive a fixed data set within a 6–month delay after the adoption of the regulation. There are approximately 1960–odd of these substances, 510 of which are petroleum products.

For the next group, we turned to those substances which are produced in quantities above 1000 tonnes but which we do not know about. We put together the annex to our regulation in a very pragmatic way – we asked Member States, "Are you aware of substances produced in quantities above 1000 tonnes?", and we then put the national lists together. We guesstimate that there are approximately 650 substances produced in quantities above 1000 tonnes which are not in the first annex to our proposed regulation and we are going to ask for a fixed data set for those chemicals within an 18-month delay of the adoption of the regulation.

Then we have another block of chemicals on the inventory which are produced in quantities greater than 10 but less than 1000 tonnes, of which we estimate there are some 8000 on the European market and for which we are asking a limited declaration form, with less information than for the first two categories, within a delay of 54 months after the entry into force of the regulation.

This is a convenient way of breaking down an enormous task. However, there is still a problem, in that when one has 2000 chemicals, as we will have in the first phase of our proposed regulation, it is not possible to deal with all 2000 of them at once.

What are we proposing to do, then? Industry and manufacturers are being asked to provide the information being requested on these chemicals directly to the Commission. We will receive this information in two parts. The first part will include the properties of the substance – physical, chemical, biological and toxicological. We hope to receive that part in a consolidated manner from industry, with only one data set per chemical.

The other part of the information is the manufacturer-related part, and this raises the question of confidentiality. This part includes the identity of the manufacturer, how much he produces and what he produces it for. Each manufacturer can submit that information independently to the Commission, but we consider that most of the information which comes under the heading "manufacturer related" will be treated as confidential.

Procedure

Let me describe how we are intending to deal with this huge mass of chemicals data. If we have 2000 chemicals, we know that we cannot work on them all at once; we have to set priorities. Every year, therefore, we intend to set a priority list of 10 to 20 substances. These priority chemicals will be farmed out to Member States, which will act as rapporteurs for those chemicals. Their job will be to carry out a detailed and in-depth risk assessment for each of these priority chemicals.

The information, decisions and recommendations from the rapporteurs will then be discussed by the Member States and the Commission together, and we will decide jointly whether there is a need for further testing, whether there is a need to reduce the risk related to a substance or whether this is already a substance of which we can confidently say, "This is a low concern substance ". If we put a substance to one side, this does not mean that we will never do anything with it – it simply means that we are temporarily putting it to one side.

Risk assessment

What I should like to do now is to try to explain what risk assessment means. Risk assessment is not an easy concept to explain, but it is essential that it be understood because it is a most important part of the chemicals control programme. It is all right to say that we will collect data and it is all right at the other end to say we want to manage or control a chemical – to have emission values to water, to control it in the workplace, to control it as a waste product – but before those management decisions can be made it is necessary to assess and evaluate the chemical.

A risk assessment, in the simplest terms, involves a comparison between the likely exposure pattern of a chemical and its effects. I think that there is no problem on the "effects" side – it is a concept which is easy to understand. Effects are basically the intrinsic properties of the chemical irrespective of how it is used – toxicity, corrosivity, carcinogenicity, does it have a propensity to explode, is it very toxic, etc. One can evaluate those intrinsic properties without any consideration of how the substance is used. This is not how a risk assessment is carried out – it is simply an identification of hazardous properties.

If, for example, a chemical is placed in a washing powder, that washing powder is going to be handled by consumers as they wash their clothes or put it into the washing machine. In addition to which, if it is a successful product, it will be sold in hundreds, if not thousands, if not ten thousands of tonnes per year. On the basis of how much of the product is placed on the market and how much of the product is used in the average load of washing done, it is possible to evaluate how much of the substance is removed in a sewage works. At the end, a calculation will provide an estimate of what the concentration will eventually be in a river. And that, in very simple terms, is a rough exposure analysis for a product used in washing powder.

The eventual environmental concentration can then be compared to existing effects data. Thus,if it is known that a product will have a very serious effect on fish with a concentration of 0.01 microgrammes per litre, and it is calculated that the average concentration in European rivers is going to be 0.02 microgrammes per litre, that substance poses a serious problem. This is a very basic way to explain risk assessment, which is a frequently used term that hides a relatively complex process because, as one can imagine, working out the average amount of a substance used in a load of washing, the average discharge to a river, the average volume of a river, etc. requires an enormous amount of effort taking into account a large number of factors to get to a predicted environmental concentration.

However, that part of the analysis can be done for any particular use pattern of a substance. For an aerosol freshener used in the home, for instance, one can calculate an exposure scenario which will give average concentration of the substance in the air in a home under normal conditions, and by comparing that to the concentrations known to be toxic or known to have deleterious effects, one has again, in very simple terms, conducted a risk assessment.

The key to the effective control of chemicals and an effective programme for the systematic evaluation for existing chemicals is to develop realistic exposure models. One can construct models that would lead to the conclusion that every substance in the world is dangerous and should be banned. This is a view that I personally cannot share. Of course, the problem is a lack of information; it is because of insufficient knowledge that exposure analysis is the weakest element in what we call risk assessment. Risk assessment will be an essential part of the existing chemicals programme.

Conclusions

To conclude, what we are trying to put together for our existing chemicals regulation is a means to collect data, a means to break the task down into discrete, smaller areas so that we can deal with them in a reasonable manner, a mechanism for bringing the data together, and a mechanism for taking collective decisions. Our hope is that we can do this in a more systematic way, before the event – before we find out that a substance will have the same impact as CFCs, before we find out that a substance should be banned like PCBs. Towards this end, we should try to develop a more predictive and systematic approach rather than the after-the-event, a posteriori approach which we have been using so far.

Discussion: Evaluation of existing chemicals

Q: The US analyses some 50 existing chemicals a year. Could the EC not have an information exchange programme to speed things up, since the Community's resources for analysis seem to be insufficient?

A: Our manpower and other resources will be greater than appears to be the case because the Member States are acting as rapporteurs. An EC–US exchange is hoped for, either bilaterally or via the OECD programme on existing chemicals.

But very few chemicals have in fact been controlled under the US TOSCA programme – they just keep accumulating further information. The EC would rather at some stage say that certain chemicals must be controlled. For the Community, the objectives of the existing chemicals programme are as follows:

- As a society, we believe a data set should exist for volume chemicals, so that we know what is in the environment;
- Rapid collection of data and identification of problems chemicals is sought. That is, action should replace classification and labelling.

Q: How will consideration of exposure patterns instead of effects affect confidentiality, since various uses will then produce various types of information?

A: There will be a manufacturer-related part of the evaluation, which will remain confidential, and a substance part, which will be consolidated by industry.

Banning and Limiting the Use of Dangerous Chemicals

by Georges Mosselmans

This type of legislation indicates that it is generally and universally accepted that the goals of protecting the health of consumers and workers, as well as the environment, are part of a single endeavour.

We know that it would be far better for this to be done in a harmonised manner at the Community level. That is why this Directive was already drawn up in 1976.

Objectives

Of course, the Directive is based on a risk assessment of the substances and preparations covered. We must ask ourselves first "What is the scope of this Directive?". First, it aims at the protection of worker and consumer health and protection of the environment. At the same time, however, it aims to prevent distortions within the Single Market as the result of differences which may exist between legislation being enforced in different Member States, for example, the use of a given substance. We know that the Treaty provides for a very high level of protection; that is the general purpose of this directive.

Scope

The Directive is a very straightforward legal instrument, comprising three articles that provide measures to restrict or ban certain substances. The Annex lists the substances covered by these measures.

The Directive is of course not applicable to transport (whatever the mode of transport used), or to substances contained in preparations that are in transit and subject to customs control. I should also mention that it does not apply to dangerous substances and preparations exported to third countries. However, I should like to note that this is not an indication that the Community is not interested in third countries – it is simply that we cannot handle a number of different things in a single directive.

Therefore, there is a Council Regulation to which we will refer later that deals with both imports into and exports from the Community to third countries.

Directive 76/769 has been amended eight times to date, but a brief summary of the content can be found in the Official Journal. I will mention a few examples of partial restrictions. Monomer vinyl chloride,for example, a substance used in the plastics industry, is covered by a total ban on its use as a propelling agent in aerosols.

Another example is tris (2,3-dibromopropyl) phosphate, which is partially banned. For example, it cannot be used as a fire retardant in clothes which come into contact with the skin, because it can cause cancer. However, this substance can still be used in floor covering (we assume that no one is going to roll around naked on the floor!).

I have taken some examples of substances which have been limited due to their dangerous effects on health. Another example is of substances which are dangerous for the environment. The organostannic compounds have been banned in any mixture which is going to be used as an anti-fouling agent in paint for boats; there are several restrictions applicable to boats under 25 metres in length. (I will explain later why this is the case.) There use is also forbidden in any appliance or equipment related to fish farming and in any partially or totally emerged apparatus.

The reason it is banned for boats under 25 metres in length is that it poses a problem of contamination of coastal waters, and it was felt that this contamination was due to great numbers of pleasure craft. It was also due to the fact that these boats were used for 2–3 months of the year at most and were usually in a port or estuary the rest of the time.

Why have we not covered large vessels as well? There are two types of large vessels – merchant vessels and military vessels. Military vessels do not, of course, fall within Community competency; this raised a problem of discrimination between merchant navy and naval vessels. In the United States, precisely as a result of this discrimination between the merchant navy and the navy, the use of organostannic compounds is still authorised. We all agree that this is a problem, but it will remain a problem as long as we are unable to deal with anything concerning the military.

Amendments

As I have mentioned, eight amendments to this Directive have been adopted. There was an amendment to the body of the Directive, which gave the Commission the right to review the substances included in the Annex and extend their restrictions, but the Council did not give the Commission the right to add new substances to the Annex.

The ninth amendment proposes a total ban on PCP, with four derogations for certain uses linked with wood treatment. One of the substances, Synogen B, was placed on the market but then voluntarily withdrawn by the manufacturer in July 1989 because it was suspected of being a carcinogen. There is thus no substitute as yet for wood treatment processes, at least as regards freshly cut wood.

At present, two Member States, Portugal and the Netherlands, are carrying out tests in this area. The Portuguese results have been quite encouraging, but for the time being at least they are not available.

The 10th amendment aims at banning cadmium. Work is being done in three sectors: pigments, stabilizers and cadmium plating. Once again, in certain cases – for example, stabilizers – we do not have 100% viable substitutes. This is why a ban on some uses has been deferred for one year after the directive enters into force. We will continue to keep a very close eye on substitute development, so that these may be used wherever and whenever possible.

In the case of cadmium, there are two types of waiver. The first involves the lights on runways; these are coloured with cadmium and will be retained for safety reasons. The second waiver involves reliability, for example the use of cadmium in computers. The available substitute is cadmium plating using silver, but this is less reliable.

A proposed 11th amendment which has been put to Council concerns a total ban on the use of PCB substitutes. It does, however, allow current transformers to remain in use because of the disposal problems that would otherwise result. The Commission is now drafting a text which will be submitted to Council in the autumn 1991. This proposes a general ban on carcinogenic, mutagenic and teratogenic substances in products intended for use by the general public. There is also a proposed 12th amendment relating to PBBEs.

To be fully comprehensive, I should mention Directive 79/117, which concerns the marketing of plant protection products – in other words, pesticides. That Directive is exactly the same as Directive 76/769. That is, Directive 76/769 covers all industrial products except pesticides, which are covered by Directive 79/117. These two Directives are aimed at harmonising Member State legislation, and the Council Regulation covers import and export at the national level.

Discussion: Banning and limiting the use of dangerous chemicals

Q: What is being done about flame retardants? The Netherlands has proposed a ban on PBBs, which are more hazardous than PBBEs.

A: A proposal on PBBs was forwarded to the Council on January 28 of this year. However, there are political and emotional, as well as environmental, factors involved in restricting PBBEs, and the former sometimes become more important than the latter.

Since 1976, balanced control has been achieved in the areas proposed, including total bans in some cases. The Commission wants responsible use – rational use – of these substances: that is, use only in those cases where they are really needed and where no safer substitute is available.

Where there is a clear danger (for example, category 1 or 2 carcinogens), a ban has been applied even without a risk assessment being carried out. But where the danger is not so clear, tests and a review of uses are needed – that is, each substance must be assessed on a case-by-case basis.

Q: PCP was banned in Denmark 14 years ago and wood preservation has continued without it. Germany and the Netherlands also have stricter proposals than the EC. Why is the Commission action proposed in the 10th and 11th amendments so weak when the Treaty calls for a high level of environmental protection, particularly as this will force the above-mentioned Member States to go to lower standards?

A: Concerning the first part of the question, I would simply like to say that the Common Position adopted by the Council on a qualified majority proposes a total ban on PCP with four exceptions restricted to professional and industrial users.

Q: And what will happen in cases of permitted use when the wood thus treated is subsequently burned?

A: Experiments showed no significant difference in dioxin produced by fresh and PCP-treated wood – the former was 2–4 ppm, the latter 3–6 ppm.

CHEMICALS IN THE WORKPLACE: A COMMUNITY STRATEGY FOR WORKER PROTECTION

by Ronald Haigh

Community directives on protection against chemicals in the workplace are of general application across the spectrum of industry, from production to placement on the market to final disposal. These are based primarily on two framework Directives - 80/1107 and 89/391.

Framework Directive 80/1107

The framework Directive 80/1107 has as its aims:

- to prevent or limit worker exposure to chemical, physical and biological agents at the place of work; and
- to protect workers who are likely to be exposed to such agents.

In the short term, Member States were required to put in place arrangements for information to workers on asbestos, cadmium, mercury and lead and to arrange health surveillance for those exposed to asbestos and lead. Soon after, specific and more detailed rules were introduced for both lead and asbestos (Directives 82/605 and 83/477, respectively).

The 1980 framework Directive listed a series of precautionary measures that were to be incorporated into any national law implementing the Directive. The main ones are identified in

Table 1. The Directive was amended in 1988 to make more specific provision for the establishment of occupational exposure limits.

The approach developed under the 1980 Directive has led to the introduction of three types of control which are likely to remain more or less intact in the future:

- for highly dangerous agents of the type our industrial society could and should dispense with: a ban on production and use (e.g. certain amines);
- for other highly dangerous agents (e.g. carcinogens) which cannot reasonably be banned: highly effective worker protection systems making use, wherever possible, of closed systems;
- for other dangerous agents: a monitoring system based, inter alia, on exposure limit values.

On the basis of the first of the above principles, Directive 88/364 was introduced in 1988 banning specified dangerous substances and certain activities associated with them. The Commission is currently preparing texts on limit values for a number of agents.

Framework Directive 89/391

The framework Directive 89/391 (Table 2) concerns the introduction of measures to encourage improvements in both the safety and the health of workers. It contains general principles for the prevention of occupational risks; the protection of both safety and health; the provision of comprehensive information, full consultation and participation of workers and their representatives, and training; as well as general principles concerning the implementation of such measures. It defines the respective roles and obligations of employers and workers in achieving these objectives and provides for the establishment of prevention, protection and emergency services at the workplace.

The measures are designed to protect workers in all undertakings (with certain specific exceptions), irrespective of the size of the undertaking. Also, the Directive covers sectors and types of work not previously covered or not covered adequately in Community rules on health and safety. The text lays down guidelines and procedural requirements for further actions.

TABLE 1

MAIN PRECAUTIONARY REQUIREMENTS UNDER DIRECTIVE 80/1107/EEC

- **Limitation of the use of the agent at the place or works**
- **Limitation of the number of workers exposed**
- **Technical preventive measures**
- **Establishment of limit values and of sampling procedures, measuring procedures and procedures for evaluating results**
- **Collective and individual protection measures, where exposure cannot be avoided by other means, as well as hygiene measures**
- **Emergency measures for abnormal exposure**
- **Information for workers**
- **Surveillance of the health of workers**

TABLE 2

SCOPE AND KEY ELEMENTS OF THE FRAMEWORK DIRECTIVE 89/391/EEC

Scope: The Directive covers all sectors of activity, both public and private, but excludes domestic servants

Duty on employers to:

- ensure health and safety of workers in every respect related to work
- develop an overall health and safety policy
- assess risks, update assessments with changing circumstances, and take preventive measures
- record risks and accidents
- inform workers and/or their representatives of risks and preventive measures taken
- consult workers and/or their representatives on all health and safety matters
- provide job-specific health and safety training
- designate workers to carry out activities related to the prevention of occupational risks
- carry out appropriate health surveillance of workers

Workers rights, responsibilities and duties:

- the right to make proposals relating to health and safety
- the right to appeal to the competent authority
- the right to stop work if in serious danger

  the responsibility for their own actions
- the duty to follow employers' instructions regarding health and safety
- the duty to report potential dangers.

Individual Directives

The first individual Directives based on Directive 89/391 have recently been adopted by the Council. Five of these Directives essentially address questions of "safety" largely neglected in the earlier texts relating to "health" concerns arising from the use of "agents" in the workplace. They cover minimum requirements for:

- the workplace itself (Directive 89/654)
- work equipment (Directive 89/655)
- personal protective equipment (Directive 89/656)
- manual handling of loads (Directive 90/269)
- display screen equipment (Directive 90/270).

In some cases, detailed technical requirements are given in Annexes or associated Recommendations. Further directives concerned with safety are set to follow. Those concerning medical assistance on board ships and on temporary and mobile worksites are at an advanced stage of discussion.

Notwithstanding the development of individual directives, the principles embodied in the 1989 framework Directive must be followed in all situations. Thus, for the future the new framework Directive will probably form the basis for most provisions on health. Indeed, new Directives concerning carcinogens (90/394) (Table 3) and biological agents (90/679) have recently been adopted as the sixth and seventh individual directives to be adopted under the new framework.

The legislation under the new framework Directive 89/391 is based on Article 118A of the Treaty establishing the European Economic Community and complements measures relating to the placing on the market of chemicals introduced on the basis of Article 100A of the Treaty.

Harmonisation of limit values

For the moment, texts relating to occupational exposure limits continue to be developed on the basis of Directive 80/1107, and a review needs to be undertaken as to how to align the two framework requirements. It is our firm intention to develop further the work on limit values, which currently provides limit

TABLE 3

THE PROTECTION OF WORKERS FROM THE RISKS RELATED TO EXPOSURE TO CARCINOGENS AT WORK

It covers exposure to:

- a substance or a preparation, labelled as R45 "may cause cancer"
- work processes at high risk, listed in Annex I

The employer must asess the nature and degree of the workers' exposure

To avoid exposure of the workers, they must:

- Either replace the carcinogens by an agent not dangerous or less dangerous
- Or use a closed production system if possible
- Or take additional measures, namely:
    - Limitation of the number of workers exposed
    - Early detection of abnormal exposures
    - Protective measures
    - Provision of information
- Health surveillance
- Observation of occupational limit values, when established

values only for vinyl chloride, lead and asbestos (Directives 78/610,82/605 and 83/477, respectively).

Some Member States establish their own limit values, whereas others frequently use the ACGIH list of threshold limit values as a guideline. In order to overcome this disparity in standards, the Community has decided to extend the adoption of Community limit values based on Directive 80/1107 as amended by 88/642. A scientific expert group has been established to provide scientific advice to the Commission towards that end.

Discussion: Chemicals in the workplace

Q: Are there any supplementary measures to fill the gaps in the existing labelling system?

A: Yes, we have proposed that the classification and labelling system should automatically be applied in the workplace. At present, the legislation is ambiguous. The measures being proposed are minimum requirements and do allow Member States to go further in the workplace, but not in marketing.

Q: It has for nine years been official Danish policy to substitute less dangerous for dangerous substances in the workplace. But changes in classification rules may decrease the possibility of classing preparations as harmful – for example, previously "irritant" preparations are no longer classified as irritants.

A (by a DG XI speaker): Classification of irritating substances has not changed. These substances are either officially in Annex I, or the manufacturer has classified them. But there are 100 substances in Denmark that are classified in ways that are not compatible with the EC Directive; therefore, labelling is no longer permitted by the Member States. If Denmark wants these substances classified, it should provide relevant information to the Commission, so that classification can be EC-wide.

A (Haigh): I would add that what I have just said applies to marketed substances. Employees and employers could get together to remove irritants from the workplace under the workplace

Directive if they wish. Article 118 of the Treaty allows Member States to maintain existing rigorous provisions in the workplace.

Q: Who is responsible for dissemination of the Directive to employers and employees?

A: People receive information differently in different countries. Dissemination is therefore more effectively done at national rather than EC level. However, during the preparation of the Directive the tripartite Advisory Committee is consulted and is responsible for making the proposals known nationally. The Directive will also exist in the form of national law and will be known as such.

Q: Is there any restriction on the information available to workers?

A: The principles and concepts of dissemination of information have been put in place by legislation, but the actual working of the process is left to individuals. Generally, however, the employee has the right to any information needed to enable him to do his job properly and safely, although the employer can then object to the parameters set, leading to compromise.

THE EXPORT AND IMPORT OF CERTAIN DANGEROUS CHEMICALS

by Alun James

This note is intended as an introduction to a Commission proposal for an amendment of Council Regulation 1734/88 concerning the export and import of certain dangerous chemicals. The most important change proposed involves the incorporation of the principle of "prior informed consent" (or PIC), which has already been adopted by the United Nations Environment Programme (UNEP) and the Food and Agriculture Organization (FAO).

Various aspects of the Regulation, including PIC, are summarised below.

Export notification

The main provision of the current Regulation, which entered into force on 22 June 1989, is a notification procedure for the export of substances which are banned or severely restricted in the Community. This procedure works as follows.

The first time that one of these substances is exported to a specific third country, the importing country is notified of the ban or severe restriction placed on it in the Community and the reasons for that ban or severe restriction. The notification is provided on an Export Notification Form which also contains references to relevant EC legislation; EC labelling requirements; details of the risks associated with the chemical; advice on the safety precautions necessary to its use; and any additional information that may be useful or of interest. Each notification form is given a reference number by the Commission which identifies the substance and the country of destination.

No Export Notification Form is required for the second and subsequent export of the substance from the EC to the same third country, but the export papers must contain the reference number allocated at the time of the first export. This is intended to help the importing country check back to the notification form obtained at the time of the first export.

Expansion of the list of banned or severely restricted chemicals

At present there are 21 chemicals which are subject to export notification. However, in the past year additional chemicals have been banned or severely restricted under amendments to Council Regulations 79/117 and 76/769. The Commission is therefore requesting that Council approve the addition of a further eight chemicals to the list subject to export notification, bringing the total to 29.

Import notification

No change is proposed in the provisions concerning import notifications. These require that Member States which receive notifications from a third country concerning the export to the Community of a potentially dangerous chemical must provide the Commission with copies of the notified information. The Commission is then required to provide a copy of the information to the other Member States and to submit appropriate proposals to Council.

Prior Informed Consent (PIC)

The introduction of a PIC scheme formed part of the Commission's original proposal for Regulation 1734/88 and was supported by the European Parliament, but this was not embodied in the original Regulation as the majority of Member States considered it advisable to await further development of PIC at the international level. However, the Council adopted a Resolution, number 88/C170/01, inviting the Commission to continue its examination of PIC and to submit appropriate proposals for the possible modification of Council Regulation 1734/88. The intent of Council to consider the possibility of introducing PIC was also mentioned in the preamble to the Regulation.

The PIC scheme now proposed by the Commission is compatible with that being set up by the International Register of Potentially Toxic Chemicals (IRPTC) on behalf of both the UNEP and FAO. This requires that a list of chemicals banned or severely restricted in various countries will be sent by the IRPTC to all countries interested in participating in the PIC scheme.

The designated authorities in these countries will be asked if they wish to ban the importation of the listed chemicals, accept the imports under certain conditions or accept the imports with no conditions. To help countries judge the hazard of the chemicals, the IRPTC will provide decision guidance documents on each of the chemicals on the PIC list.

To date, over 80 countries, including the USA and Japan, have notified the IRPTC of their interest in participating in the voluntary PIC scheme set up by UNEP and FAO. The Commission considers that the introduction of a regulation making the same PIC scheme mandatory in the Community would be a positive development in the international control of dangerous chemicals and would have beneficial effects on human health and the environment in third countries. All that would be required of exporting industries in the Community would be to comply with the wishes of participating third countries on the import of dangerous chemicals.

Discussion: Export and import of hazardous chemicals

Q: Are all produced but not-registered chemicals included in the PIC scheme?

A: Chemicals are listed as Annex I substances by being either restricted or banned in other Directives; Council then reviews that severe restriction list. Thus, a chemical that has never been registered will probably not be banned or restricted.

Q: Will PIC now be accepted by Council, which rejected it in 1988?

A: In 1988, some Member States feared a competitive disadvantage in relation to the US and Japan. Now, those two countries are also joining PIC, so this is no longer a problem.

Q: Has any progress been made on Member State reporting of information on notifications, as provided by Article 7?

A: A database is being circulated among Member States, and we hope that sufficient data will now be available to compile reports.

The control of ozone-depleting substances

by George Strongylis

Scientific warnings throughout the 1970s and 1980s have recognised that ozone depletion is a global problem requiring a global solution. Scientists saw their most pessimistic predictions confirmed with the discovery in 1985 of enormous ozone reductions, averaging about 50% in the spring Antarctic ozone. The simultaneous measurement of levels of chlorine some 50 times higher than their natural background proved that the agent responsible for this depletion was chlorine-containing man-made chemicals.

Additional damaging evidence comes from ground-based observational data which show that a 2–3% reduction in global ozone occurred between 1959 and 1986, surpassing the 1% depletion predicted by theoretical atmospheric models.

Ozone-depleting substances

The main substances causing these effects are chlorofluorocarbons (CFCs), halons and, as we have recently come to realise, substances such as 1,1,1-trichloroethane and carbon tetrachloride. The chlorine and bromine that these substances contain interact with a number of chemical components in the atmosphere to produce ozone depletion through a complex series of more than 100 chemical processes. It is a very complex phenomenon we are addressing.

Another important problem relating to these compounds is that they make an appreciable contribution to climate warming through the greenhouse effect. CFCs are estimated to account for some 15–20% of global warming.

It is important to stress the long-term character of these problems. CFCs have a lifetime in the atmosphere of 50–140 years, depending on type. It takes them about 15 years to penetrate into the stratosphere; thus, we have yet to see the destructive effect of the large quantities of CFCs used over the last two decades. Even urgent action to limit CFCs will therefore not return ozone to its normal levels before 2007.

We are dealing with a serious ecological problem which confronts us with a new feature – that of inter-generational responsibility. To what extent will our actions affect the lives of our children and beyond? This consideration is of overriding concern in this issue.

About one million tonnes of CFCs are currently produced in the world every year, of which some 40% are produced in Western Europe, 30% in North America and 20% in Asia and the Pacific. Owing to their non-toxic character, CFCs have become ubiquitous components of our consumer society. They are widely used in aerosol cans (although that application has declined significantly), in refrigeration, foam plastics for furniture and insulation and in cleaning electronic components.

International measures

Attempts by some countries to ban their use in aerosols were followed in 1985 by the Vienna Convention for the Protection of the Ozone Layer. The Vienna Convention, while including cooperation on research and information exchange, did not succeed in instituting control measures. These came in 1987, with the conclusion of the Montreal Protocol dealing with substances that deplete the ozone layer. The Protocol is an achievement of international cooperation for the protection of the environment. It establishes a unique set of controls on the production of chemicals, instituted internationally for environmental reasons.

The Protocol has another feature. It recognises the special situation of developing countries and gives them a grace period of 10 years' exemption from these controls if they consume under 0.3 kilogrammes of CFCs per capita yearly. (It can be mentioned indicatively that in the European Community and in the US in 1988 we consumed about 0.85 kilogrammes per capita.)

This treatment of developing countries stems of course from the recognition that their development should not suffer during the difficult transition to substitutes for CFCs that will undoubtedly cost more and will, at least initially, be the privilege of the developed world. These provisions are meant to encourage a sense of solidarity in the developed countries towards the developing world, in the context of a common threat.

Community measures

At Community level, the control measures set out in the first version of the Protocol are implemented by Council Regulation 3322/88 entitled 'On certain CFCs and halons which deplete the ozone layer.'

At their second meeting in June 1990, in London, the parties to the Montreal Protocol agreed to tighten up the Protocol because they had, since its conclusion, received very bad news concerning severe depletion above Antarctica and the general effects that this could have on the earth's atmospheric balance. In particular, they decided that CFCs and halons should be eliminated earlier than originally foreseen and that control measures should be extended to other chemicals dangerous to the ozone – namely, the other fully halogenated CFCs, carbon tetrachloride and 1,1,1-trichloroethane.

The parties also decided to establish a financial mechanism for the three-year interim period 1991–93 and beyond, to help developing countries meet the requirements of the Protocol. Nourished by developed countries with some US$ 240 million in addition to existing transfers, this mechanism is an important incentive for developing countries to accede to the Protocol.

Indeed, because of their high potential consumption of ozone-depleting substances, the accession of developing countries to the Protocol is the key factor in the success of all the endeavours undertaken so far. The mechanism is implemented under the auspices of the executive committee of the fund, which has its seat in Montreal and is composed of seven developed and seven developing countries. The committee has met twice so far and has already made some progress in solving the issues related to the implementation of the mechanism.

At Community level, the revision of the Protocol is being implemented through a new Council regulation which replaces the old one, 3322/88. The number of this new regulation – 594/91 – has just been published in the Official Journal.

In line with the London review of the Protocol, the EC Environment Council approved this regulation at its session of 20–21 December 1990; it decided progressively to reduce and finally to prohibit the production and consumption of all substances covered by the new Protocol. However, following a Commission proposal, the Council has been more ambitious than other parties to the Protocol. Thus, the new regulation foresees a phase-out schedule that is more stringent than that decided in London. For example, chlorofluorocarbons and carbon tetrachloride would be eliminated by June 1997 in the Community, while the other Protocol parties, particularly our major competitors – the US, Japan and the Soviet Union – plan to eliminate these substances by the year 2000.

The Commission also intends to implement in the near future the London resolution on the so-called transitional substances which are coming on to the market to replace CFCs. The Commission wishes to create a legal framework by which CFC substitutes (HCFCs) will be phased in, used responsibly for a time and then phased out.

These potential substitutes would be able to solve about 95% of the ozone depletion problem, with low negative impact on the ozone layer. Their utilisation should be controlled and limited in time, until completely clean substitutes are available.

Voluntary agreements

We have also in the past few years undertaken a programme of voluntary agreements with industry. These were meant to bridge the gap between the first version of the Protocol and the new version, which provides for much more drastic action, since we quickly realised that the former was not stringent enough to lead to replacement of the ozone layer.

Those agreements have proved very useful in providing a basis for serious discussion with industry to assess the real possibilities for reductions. They resulted in three Commission recommendations, in three different industrial sectors with traditionally heavy use of CFCs – aerosols, refrigeration and foam plastics. The recommendations foresee even more drastic cuts in CFC consumption than those instituted by the new regulation – namely, a 90% reduction of CFCs in aerosols by 1990 (this has already taken place), a 65% reduction by 1993 in foam plastics and a 50% reduction of CFCs in refrigeration.

Conclusions

The combination of all these measures gives the European Community a leadership role in the protection of stratospheric ozone. We intend to continue in this role. A new revision of the Protocol is foreseen for 1992, based on a request made by the European Community at the London meeting in June 1990. We are not happy with the position of the United States and other countries which have not agreed to our proposal for a 1997 phaseout, and we intend to pursue this matter further in 1992.

We are proud of our record, and we believe that other countries will be joining us as the result of the appearance of new substances on the market. We cannot forget, however, that the scientific evidence tells us we should be phasing out even more rapidly. Consequently, we will remain vigilant and do everything we can – for example, recycling – to encourage the lowest possible production of this substance in the years to come.

Discussion: The control of ozone-depleting substances

Q: What is the EC policy on the phase-out of HCFCs?

A: That is being discussed now. The Protocol gives the phase-out date as between 2020–2040, but the EC is not in agreement with this. We take a more pragmatic view and would like to come to an agreement with industry on how to minimise HCFC use and phase them out as soon as possible. We are hoping to develop this policy during the coming year.

Q: DuPont, ICI and other producers are providing data on the environmental effects of "soft" substitutes. Is the EC content with the information available, or would it like more control over this area?

A: DG XII – research – has conducted and funded ozone research for over 10 years. It coordinates research at the European level and provides advice to DG XI on how to deal with HCFCs. Thus, DG XI in fact obtains its information from scientists. Companies for their part provide information on use, production methods, etc.

Q: How will the EC control the import of CFC-containing goods produced outside the Community?

A: The Protocol does not control this either – it only controls bulk trade. However, if global production is controlled and participation in the Protocol is high, the rest will be controlled automatically. That is the philosophy of the Protocol.

Nonetheless, the UN is at present examining the subject of international product control, despite the awareness that this is a very difficult area to control.

Regulation of Medicinal Products in the European Community

by Fernand Sauer

Pharmaceuticals are not really chemicals in the ordinary sense. First of all, many pharmaceuticals which are at present on the market are based on natural products; more than one third are based on substances which are not well defined from the chemical point of view, many of biological or vegetable origin. With more advanced technology, we will of course be seeing still more of these.

Secondly, the pharmaceutical sector is a particularly difficult sector to harmonise at the European Community level. It is consequently a combination of three national policies: public health policy, industrial policy and social policy, and it is very difficult to get Member States to abandon sovereignty in these areas. The whole process of pharmaceutical legislation has therefore been a long and very slow one.

Historical background

The major landmarks are 1965, with the thalidomide case, which led to the introduction in all Member States of the basic principle that any new pharmaceutical introduced to the market would have to undergo pre-marketing testing. The principles of quality, safety and efficacy were adopted in 1965, but we needed 10 years – until 1975 – to qualify the actual rules which were to apply to the testing of drugs.

The first harmonisation directives were adopted in 1975, and the main committee – the Committee on Proprietary Medicinal Products (CPMP) – was established in order to coordinate national decisions taken in the field of pharmaceutical distribution.

In 1981, the Community adopted similar provisions for veterinary medical products. The system was reformed in 1983. In 1987, a series of biotech and high-tech products was taken into consideration; and in 1990, 11 proposals were submitted to the Council and to Parliament in order to achieve the harmonisation needed to complete the Single European Market by 1992.

Summary of legislation

First of all, we have republished all applicable texts in five volumes in a series called "Deadline '92 – A Frontier-Free Europe". Volume I deals with the rules applicable to human medicine. Volume II is called "A Notice to Applicants" and contains a standardised format for the presentation throughout Europe of marketing authorisation dossiers. These extensive dossiers must be sent to all 12 Member States; we have harmonised the presentation standards for these dossiers and agreed on these standards with the EFTA countries as well. We thus have one standard and a simplified technology accepted throughout Europe.

In addition to harmonised testing, we have also introduced various principles for certification and inspection similar to those applicable in the chemicals sector; for example, the GLP (good laboratory practices) rules are the same, and there are cross references to the GLP rules for chemicals. There is a long tradition of good manufacturing practices in the pharmaceuticals sector, and we have a series of texts to ensure the highest possible quality of manufacture.

As of 1992, these principles will apply not only to pharmaceuticals intended for intra-Community trade, but to exports as well. The same standards will thus apply to products for export as to products consumed within the Community.

More recently, we have introduced stringent principles for the control of clinical trials; these are called Good Clinical Practices and are specific to the pharmaceutical sector.

Table 1

The Rules Governing Medicinal Products in the European Community

VOLUME I	The rules governing medicinal products for human use in the European Community Catalogue number CB - 55 - 89 - 706 - EN - C
VOLUME II	Notice to applicants for marketing authorizations for medicinal products for human use in the Member States of the European Community Catalogue number CB - 55 - 89 - 293 - EN - C
VOLUME III	Guidelines on the Quality, Safety and Efficacy of medicinal products for human use Catalogue number CB - 55 - 89 - 843 - EN - C ADDENDUM : CB - 59 - 90 - 936 - EN - C
VOLUME IV	Guide to Good Manufacturing Practice for the manufacture of medicinal products Catalogue number CB - 55 - 89 - 722 - EN - C
VOLUME V	The rules governing medicinal products for veterinary use in the European Community Catalogue number CB - 55 - 89 - 972 - EN - C

The so-called "extension directives" covering GMP (Directive 89/341/EEC), vaccines (Directive 89/342/EEC), radiopharmaceuticals (Direc- tive 89/343/EEC) were published in the Official Journal Nr. L 142 of 25.5.89, and the Directive on blood products (89/381/EEC) in O.J. L 181 of 28.6.89.

N.B. These texts are on sale at the :

Office for Official Publications of the European Communities
2, rue Mercier
L-2985 Luxembourg
Tel. (352) 49 92 81
Telex PUBOF LU 1324 b

In recent years, the Council of Ministers has given us a mandate to extend pharmaceutical administration to many categories which were not previously covered – in particular, vaccines, blood products and radio-pharmaceuticals. These texts were adopted in 1989 and will come into force in 1992.

The Council has also recently adopted a series of texts for updating legislation relating to veterinary medicinal products; specifically, the Council for the first time adopted a regulation in our sector which set limits for residues of veterinary medical products in food.

Procedure

The whole system of pharmaceutical administration is based on the functioning of many committees and expert groups: a total of 400 experts from governments and universities help us in shaping EC legislation and advise us on the opinions we have to prepare. The Committee on Proprietary Medicinal Products has several working parties on quality safety efficacy.

There is a special group on pharmaco-vigilance, which means post-marketing surveillance of the side effects of drugs. There is also a group on biotechnology in the pharmaceuticals sector, including inspectors, committees on pricing, social security and veterinary medicine products and, finally, a pharmaceutical committee which is a meeting of the Directors General of Pharmacy of the 12 Member States. They advise us on general pharmaceutical policy.

Coordination procedures

At present, registration of medicinal products is harmonised from the legislative point of view but decisions are made by each Member State. Nevertheless, there are several coordination procedures. Firstly, there is a coordination procedure for biotech products. Before any biotech medicinal product is put on the market, it must undergo a Community evaluation. This is initiated by one of the Member States acting as rapporteur, and the Committee reaches an opinion which is then transmitted to the 12 Member States and to the Commission.

TABLE 2

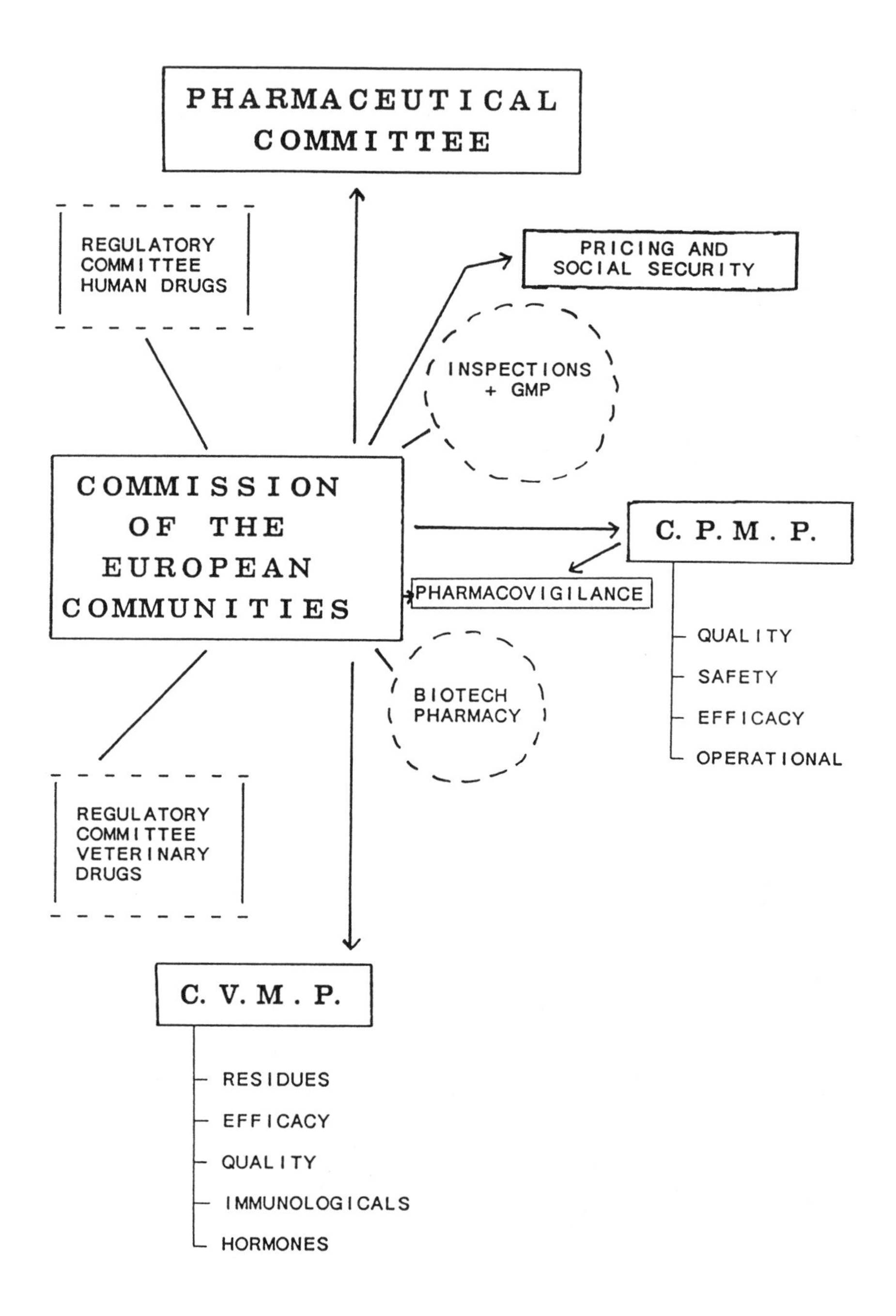
PHARMACEUTICAL COMMITTEE
REGULATORY COMMITTEE HUMAN DRUGS
PRICING AND SOCIAL SECURITY
INSPECTIONS + GMP
COMMISSION OF THE EUROPEAN COMMUNITIES
C. P. M. P.
PHARMACOVIGILANCE
QUALITY
SAFETY
EFFICACY
OPERATIONAL
BIOTECH PHARMACY
REGULATORY COMMITTEE VETERINARY DRUGS
C. V. M. P.
RESIDUES
EFFICACY
QUALITY
IMMUNOLOGICALS
HORMONES

The second procedure, applicable to conventional drugs, is called the "multi-state procedure", by which a company may require Member States to recognise an initial evaluation done in one Member State.

The third coordination mechanism relates to side effects and pharmaco-vigilance, where we can, at the initiative of the Commission or one of the Member States, agree on special measures either to withdraw products from the market or to limit the conditions of use.

The outcome of these procedures was, briefly, as follows. We originally had very few applications. The system is totally voluntary except for biotech products, and from 1978 to 1986 we had only 41 application dossiers. More recently, in only three years, we have had over 124 dossiers submitted.

For multi-state applications, where one country evaluates and others are asked to recognise that evaluation, the system has become much more effective in recent years. In 90% of cases, opinion is now favourable to the evaluation of the first country, which was not the case before.

This demonstrates that we have achieved a degree of harmonisation. However, in our judgement, this experience is not sufficient to allow for automatic mutual recognition of decisions. For the high-tech, biotech procedure, more than 30 dossiers have now been submitted, and in all cases opinions have so far been favourable.

Economic aspects

Besides public health, we have also had to tackle the economic aspects of the marketing of products. Within the Community, there are enormous price differences for medicinal products; these may range from 100–300% for the average price in Spain, for example, as compared to that in Germany.

There are also major differences in consumption, which are not always paralleled by differences in price. In countries such as France, for example, prices are apparently very low, but consumption is so high that the social security expenditure is similar to the very high levels in Germany.

TABLE 3

OUTCOME OF COMMUNITY PROCEDURES

TYPE OF PROCEDURE	FORMER CPMP (75/319/EEC)	MULTI-STATE (83/570/EEC)	HIGH/BIOTECH (87/22/EEC)
PERIOD	**1978–1986**	**1986–DEC.'90**	**1987–DEC.'90**
NUMBER OF DOSSIERS	41	149	33
NUMBER OF APPLICATIONS	253	784	342
TOTAL NUMBER (OPINIONS)	41	95	18
- FAVOURABLE	28	87	18
- UNFAVOURABLE	13	9	0
SUBSEQUENT NATIONAL DECISIONS			
AUTHORIZATIONS	175	312	83
REFUSALS	63	61	0
OUTSTANDING	15	118	36

It is not envisaged that all state social security regulations be harmonised in the present situation. Nevertheless, a first directive on the transparency of pharmaceutical pricing entered into force last year. This directive did not fundamentally change the rules of control of pricing of social security reimbursement but has adapted systems which are very different to allow for the free movement of medicinal products in the Community.

We are now, on the basis of experience with this directive, about to propose a further step in that direction, before the end of the year.

European drug data bank

At the same time, the European Parliament has asked the Community to begin work on the creation of a European drug data bank, which would be open to both the general public and the professional public. This drug data bank would contain two sets of information – one, the scientific information on conditions of use in each country; and the other, prices and the costs of daily treatment. This would cover products present on a majority of Member State markets. In this way, even if there are hundreds of thousands of products on the market, the 2000–3000 products included in the European drug data bank would represent over three quarters of Community consumption.

Pharmaceutical patents

One of the proposals made in that respect, which were hotly debated in Parliament last year and are now being debated in Council, is a suggestion that there should be restoration of patent terms in the pharmaceuticals sector. Long tests are performed on pharmaceutical products, and real patent life is extremely short – some 8 years instead of the 20 years customary in other sectors. If adopted, this regulation would only partially resolve this problem, by extending the patent term to 16 years of effective patent life.

Rational use of drugs and advertising

At present, we have three proposals in Council, which will be discussed by Parliament in May, with the purpose of encouraging a more rational use of drugs in the Community. Formerly, we concentrated on the product itself; we have now started concentrating on the software which goes with the product – that is, information, prescription rules, self-medication, etc.

Another very controversial subject being dealt with in Parliament and Council at present is a proposal on pharmaceutical advertising. Two proposals on alternative medicines, in particular homeopathic medicines, are also before Parliament and the Council at the moment.

Future system of registration in Europe

In addition to the seven other proposals currently before Parliament, there are four proposals for reforming the registration system in the Community – the major subject of this year. These envisage the creation of a European medicines evaluation agency, similar to the idea of the environment agency but with a certain degree of responsibility for decision-making. The idea is based on experience with current CPMP procedure – multi-state on the one hand, which tries to obtain a sort of mutual recognition from Member States, and biotech on the other hand, which is a more centralised procedure.

We have retained the two largely to provide a choice for companies – there are small companies which cannot afford to go central, but there are other companies who prefer to go central immediately. There is a large choice open to companies.

The system which is currently more often used is the decentralised system. When a Member State carries out an evaluation, it must be recognised; if it is not, there will be binding arbitration through the agency. Secondly, there is the centralised registration procedure, which is obligatory for biotech products, enhancers in the veterinary field, and any new drug or chemical entity. Unity of Community decisions is important in these areas.

In this case, the evaluation will be carried out by the agency and finally endorsed by Community institutions through a marketing organisation valid throughout the 12 Member States. The national authorities thus have no say in this central authorisation system.

European medicines evaluation

The structure envisaged, which is currently under discussion, is in fact not very centralised. Last year we discussed the possibility of a European food and drug administration similar to the US Food and Drug Administration; however, considering the bureaucratic tendency of the US agency and the impossibility within the Community of dismantling the 12 national systems and replacing them with an institution of 8000 officials, we made a different choice.

Our plan is to create a permanent secretariat of about 100 staff to evaluate dossiers, write reports and so on, but the main evaluation will be carried out by consultants from Member States. Experts from universities and national regulatory authorities will serve both the agency and the national authority.

If there is a continual exchange of experience, consumers will continue to be as well protected as they are at present, and there will be a plus in the form of general familiarity with the field. The present committees will be transformed and adapted to the future system but will basically remain the same committees and expert groups with the addition of a much larger volume of outside expertise flowing into the system.

As in the case for the environment agency, the location of this agency remains a problem, and no forecast can be made at the present time.

The proposal for the agency will go to Parliament for a first reading in June or July. It should be noted that the agency is not a Commission service. It is outside the Commission and comprises a sort of partnership between the 12 ministries of health and the Commission. Scientific evaluation will come from the agency, with leaflets bearing labelling in all nine Community languages; it will be up to the Commission to transform these opinions into binding decisions.

Safeguard clause

There is a safeguard clause that allows any Member State to oppose a Commission decision. In such a case, the matter is referred to a regulatory committee where the Member State can oppose the Commission proposal. If there is still no agreement between the Commission and the majority of Member States, the matter then goes to the Council for a final decision. It does not automatically go to either the Council or the regulatory committee because these decisions are basically scientific and should not be made political.

International harmonisation

This harmonisation process takes place against a background of other international action. Many years ago, we had already initiated cooperation within the framework of the Council of Europe. There is also a European Pharmacopoeia, which defines for EFTA and the EC countries the quality standards of active substances, and the EC will join this Pharmacopoeia on behalf of its 12 Member States, probably next year.

We are actively negotiating with our EFTA partners on various issues, but these discussions will probably be overtaken by the general treaty which could be signed in June between the EC and EFTA. We also have regular contacts with the US Food and Drug Administration and with the Japanese Ministry of Health in order to obtain better international harmonisation procedures.

In that respect, we have agreed with the US Food and Drug Administration to hold a first international symposium in Brussels in November of this year in order to initiate a five-year programme for the (if possible) total harmonisation of pharmaceutical requirements throughout the world. This should also have a positive effect in reducing the development costs for drugs.

Discussion: Regulation of medicinal products

Q: What specifically are the links between the Community and the Council of Europe in this area? When one member state has given approval, will this be accepted in all member states?

A: Generally, the Council of Europe tends to adopt non-binding decisions. The Pharmacopoeia is compulsory. Nevertheless, the decision to register a finished product is to a certain degree discretionary and depends on a risk/benefit analysis. Consequently, there is no automatic acceptance by member states.

Q: Will it be possible to have a requirement that standard warnings appear on pharmaceuticals?

A: The labelling reform that has been undertaken is being carried out in close cooperation with consumer groups. The current idea is to make labelling as informative and legible as possible.

Pregnancy warning guidelines are being worked on, but these are complicated by product liability rules. Companies are for their part content to exonerate themselves from all liability by letting the Member States rule out all such use, which would leave pregnant women without any treatment.

Q: The four proposals linked to the proposed agency have no medical ethics aspect.

A: Ethical considerations are a constant preoccupation. To some degree, these are covered in the Helsinki principles and also in our EEC Good Clinical Practices. In addition, a type of external audit will occur in the form of a scientific and ethical High Council.

However, there are other problems, such as those linked to contraception. The provisions of the future system are meant to respect such cultural differences between the Member States.

Food legislation and food additives

by Willem Penning

It is, in fact, very difficult to present a complete overview of what is happening in the field of foodstuffs in a short space. Many things are going on in this area; we are working on a wide variety of directives, and framework directives in fact already exist.

In 1962 the Commission already had its first directive on chemicals which are used in food for a technological purpose. The first directive on colouring matters in food was also issued in 1962. The directive was amended a number of times – up to eight times – in order to restrict the use of certain colouring materials. In fact, some colouring materials were deleted because they were not up to the new scientifically determined standards which had been adopted over the past years. Several other directives were then put forward, dealing with anti-oxidants, emulsifiers, thickening agents, and so on.

Objectives

The basic principle of all these directives is that they lay down a positive list – which means that only these food additives may be used, to the exclusion of all others. Introduction of another additive is therefore completely unlawful. For example, EDTA, which is a complexing agent, may no longer be used because it is not taken up in our directives, although some Member States continue to use it.

These directives have very rarely laid down rules about where these additives can be used. Therefore, in 1989, the Commission proposed a framework directive in which the principle rules are laid down by which one cannot state that such a food additive can be used as a food additive. To use a food additive, one must first demonstrate technological need; secondly, the additive should not induce any unhealthy condition – it must be toxicologically safe. Thirdly, it is not permissible to induce the consumer into error. For example, introducing a colouring agent so it looks just like another product should be considered unlawful.

The framework directive on sweeteners requests that the Commission propose implementing directives in which positive lists of additives and foodstuffs are laid down. The various categories of food additives are shown in Fig. 1. These obviously include colours, preservatives and anti-oxidants, but further down the list of additives laid down in the framework directive are also sweeteners, bulking agents, and so on.

Technological need

How do we prepare such directives on food additives? It is not a particularly easy task. First, we must be sure that a technological need is present. If, for example, sorbic acid is put into a medium which is not acid, it will not be effective; in this case, the use of sorbic acid can clearly be regarded as unnecessary and authorisation should therefore not be granted.

To determine technological need, the Commission must obtain information from industry. It is quite clear that food technologists only can be found in this position. We are thus asking the industrialists who prepare our foodstuffs to establish a technological need. When an industry is unable to demonstrate this technological need, we find it difficult to see that a food additive need be used in that particular product.

The most difficult thing for us is that we in the Commission ought to be aware of all those foods in which additives are used. However, in difficult cases, the Commission may prepare a directive and miss out some categories of food because it was simply not aware of them. This is the case for the proposed directive on sweeteners which is now before Parliament, the Economic and Social Committee and the Council in the cooperation procedure: we could not have imagined, for example, that sweeteners such as saccharine are used in snack-food – not for their sweetening power but for flavour enhancement.

CATEGORIES OF FOOD ADDITIVES

- Colour
- Preservative
- Anti-oxidant
- Emulsifier
- Emulsifying salt
- Thickener
- Gelling agent
- Stabilizer
- Flavour enhancer
- Acid
- Acidity regulator
- Anti-caking agent
- Modified starch
- Sweetener
- Raising agent
- Anti-foaming agent
- Glazing agent
- Flour treatment agent
- Firming agent
- Humectant
- Sequestrant Enzyme
- Bulking agent
- Propellent gas and packaging gas

Acceptable Daily Intake

What are the principle rules which guide us in allowing certain amounts of additives in certain foods? It is the ADI – the Acceptable Daily Intake. This concept is now accepted worldwide; it is used in considering provisions for a proposal not only by the Commission, but also by JECFA – the Joint Committee for Food Additives and Contaminants – which meets every two years in Geneva.

ADI is expressed in milligrammes per kilogramme of body weight. For saccharine, for example, the ADI is 2.5 mg per kg body weight. A person can then multiply that by his own body weight to know how many milligrammes of that substance may be taken every day without leading to difficulties. This is a concept which can be discussed, but it is the only one we have at the moment and is derived from animal studies.

The Scientific Committee for Food

The Commission has no competence to evaluate data, and it would be very difficult for a legal organisation to make scientific evaluations. Therefore, in 1974, the Commission set up a scientific committee for food which has to be consulted on all questions that have an impact on human health.

How does this committee work and who sits on it? It includes no officials of the European Community. It is composed of university professors from the Member States with a high degree of competence in matters relating to food: this includes nutritionists, toxicologists, medical doctors and specialised nutritionists (small children, babies). There are 18 members and five counsellors. These people meet up to 12 times a year and hand down opinions on all matters put to them by the Commission.

They deal with a wide variety of subjects. These include RDAs (Recommended Daily Allowances), vitamins, general criteria for the evaluation of food additives, and natural substances. The last category may pose considerable difficulty: for example, deciding whether the steady increase in mushroom consumption over the past five years is now beginning to pose a danger to human health because of the presence of toxins in all mushrooms.

The scientific committee does not have sufficient time to deal with these matters in plenary session. It is therefore divided into several sub-groups – we call them working groups – to which other competent people may be invited to give their opinions. Opinions of the Scientific Committee for Food, which is an advisory committee and in no way a legal committee, are always unanimous. The Committee publishes its reports in booklets which can easily be obtained from the Community publications office in Luxembourg.

Establishing a dossier

It is normally industrialists who present dossiers for evaluation. DG III/C/1, foodstuffs division, has published a volume entitled "Presentation for an application for the assessment of food additives". This publication describes fully how the dossier, which should be sent to the Commission services in 35 copies, should be built up. A wide variety of information is requested. It goes from analytical chemistry up to complex toxicology, including long-term studies, carcinogenic studies, long-term feeding studies, etc., as well as toxicology, acute toxicology, mutagenicity and short-term testing.

We are recommending to those presenting dossiers for evaluation of new chemicals or new additives that they follow the directive issued by Directorate General XI on good laboratory practice. Dossiers that do not comply with that directive will not be accepted; the industry will be asked to withdraw its dossier or carry out further studies, if the Committee so wishes. The Committee will express an opinion, such as that the food additive is acceptable, or is acceptable under certain conditions, which is the acceptable daily intake, or the Committee may even go so far as to say that an additive is only acceptable in a particular food class where there is a technological need and where the ADI would not be exceeded.

We occupy a very difficult position at the moment. The framework Directive on food additives states that all food additives should pass through this procedure. However, Member States normally have the same sort of national system for evaluation; this leads to a situation where certain countries are continuing their own evaluation, producing their own ADI and thereby creating difficulties for the Scientific Committee for Food.

The Commission has taken a very strict position on this. It is only the opinion of the Scientific Committee for Food which will form the basis for laying down a directive regarding limitations on the use of additives in foodstuffs.

The difficulty of such a situation is illustrated by the sweetener cyclamate. As mentioned earlier, the Commission has been proposing a directive on sweeteners which includes the use of cyclamates. Cyclamate has received approval from the Scientific Committee for Food, but the CUT in the United Kingdom has now issued a press release in which the Minister calls cyclamates unacceptable.

The Scientific Committee for Food is continuously monitoring food additives. When a dossier has been evaluated and an opinion has been expressed, that opinion is not immutable. The Scientific Committee for Food first looks into the problem of whether consumers are exceeding the ADI, which involves continuous monitoring; secondly, new scientific information may create the need to review the ADI itself because of new scientific developments.

Member State coordination

The Scientific Committee for Food, unlike the Commission, now has the power to make an overview of what exactly is being consumed in Member States. This is a very difficult area, because as one goes from the north of Denmark to the south of Greece, the consumption pattern of foods changes drastically; nor is it particularly easy to determine what quantities are used in any one country. Denmark, for instance, consumes a lot of milk products in which certain additives are present while this is not the case in Greece.

The Commission has in recent years been proposing to the Council a directive to be adopted under Article 100A (procedural cooperation between Parliament and Council) which establishes a cooperation system between Member States and the Scientific Committee for Food. In the preceding paper, Mr Sauer described an agency of a sort, although the food division would not see this new system of cooperation between Member States as an agency. It will consist of the Commission and its staff in collaboration with Member States, which would carry out the prior evaluation of dossiers and deal with questions about foodstuffs in general – not merely additives –after which the Scientific Committee for Food with its 18 members would finally deliberate the matter.

The Directorate General III considers that for carrying out such an exercise, the Commission needs about 15 staff in Brussels, and a regular staff in the Member States dedicated specifically to these tasks. Their work will range from research to active laboratory research when there is something that needs to be controlled, the formation of opinions and drafting reports for the Scientific Committee for Food, which will then decide whether food additives or any other food matter are acceptable for human consumption.

Conclusions

A final word about the current situation for food additives. The Commission has issued a proposal on sweeteners. It intends to issue a directive on colouring materials and miscellaneous additives by June, and a further directive on all other food additives by November of this year, so as to put us in a good position before the completion of the internal market in 1992.

The reason for all this effort in putting together common rules for the European Community is, first, the protection of human health and, secondly, eliminating the problems and trade barriers which still exist in Member States.

The Scientific Committee for Food is now evaluating guidelines on food products prepared using biotechnology. We have drawn a lesson from international experiences with tryptophan. Prepared biotechnologically in Japan, this substance has proved to contain a highly toxic compound; it has also led to several fatalities in the United States.

The Scientific Committee for Food will thus look into defining general criteria for the evaluation of products prepared by biotechnology. If we limit ourselves to dossiers on simple chemicals, we cannot fulfil our duty under Article 36 of the

Treaty, which provides that all acts of the Community should provide a high level of protection where health is implicated.

We are also preparing directives on hygiene and on novel foods. There also, the Scientific Committee for Food will provide us with general criteria. How the directive on novel foods is going to look is far from clear, though ideas will certainly come up before the end of the year.

Another fairly big task assigned to us is a regulation on flavouring. We have been asking industry to draw up an inventory of all flavourings from both natural and synthetic sources. We now have an inventory of some 18 000 items, and the Scientific Committee for Food must now tackle the problem of determining allowable quantities.

Discussion: Food legislation and food additives

Q: Labels do not indicate the more subtle effects of food additives, such as allergic effects due to colouring agents. Is the EC making any provisions to meet this problem?

A: Allergic reactions are a problem, but their basis is completely different from that of toxicity. Toxicity is the effect an additive has on the body; allergy is the body's response to an additive. It is thus highly individual.

A balance must be sought between technological need and adverse individual reactions. This might eventually be resolved through labelling. At present, E numbers are a guide to personal allergic problems.

Q: The EC Directive states that colouring is to be used only when it is technologically necessary. When is it really necessary to use colouring in basic foods?

A: This point has been debated for a long time with the Member States. Colouring is used to make foodstuffs attractive. Other uses are to provide uniformity, to make food more acceptable for consumption, and to restore colour lost in processing. It is difficult to justify these as technological need. However, the Member States will not agree to rule out all colours.

NEW LEGISLATION ON THE AUTHORISATION OF PLANT PROTECTION PRODUCTS

by Alberik Scharpe

My intention is to explain the background to the new legislation concerning authorisation of plant protection products, proposed in 1989 by the Commission (COM(89) 34) and currently in a far advanced state of discussion in Council, particularly since the European Parliament delivered its opinion on this proposal in February. I shall also outline its main provisions and the functioning of the proposed Community registration system, as well as its status within the whole framework of Community legislation relating to pesticides.

Background

As a first element of background, I would mention the 1988 Communication from the Commission to Council, "Environment and Agriculture", in which the general policy of the Commission with regard to pesticides was set out.

In this Communication the Commission recognised:

- increasing concern about the environmental impact of pesticide use;
- the disappearance of the agricultural deficiency of the early years of the Community and a growing interest in the de-intensification of Community agriculture and in forms of alternative agriculture.

In this Communication, the Commission indicated that the general objective of Community policy would be to reduce to a strict minimum the use of pesticides and discussed a range of possible measures it might propose in the future. The

Commission announced the Directive on authorisation of pesticides as a first step and, beyond this, other measures, some of which have also been developed since then and should briefly be mentioned in this context:

- a proposal for a Council regulation on the protection of organic farming (COM(89) 552);
- a proposal for a Council regulation on the introduction and maintenance of agricultural production methods compatible with the requirements of the protection of the environment and the maintenance of the countryside (COM(90) 366).

Secondly, the new legislation on the authorisation of plant protection products falls within the about 300 measures proposed in the so-called "White Paper" and necessary to complete the internal market by 1992 along the principles of the Single European Act of 1986, calling for the creation of an internal market without any internal frontiers. The realisation of this objective in the area of plant protection products inevitably requires a considerable effort of harmonisation, which is reflected in the proposal we have under discussion.

Harmonisation of plant protection products registration

The basic problem to be solved in respect to harmonisation is how to reconcile the objective of creating conditions under which plant protection products may circulate in the Community on the one hand, with the need to ensure a high level of protection for human and animal health and for the environment on the other hand.

This problem is posed in particular for plant protection products, of which the performance and even aspects of safety and environmental impact may depend on local or regional agricultural, ecological and plant health conditions.

Plant protection products are therefore prepared under formulations which are finely tuned to the agricultural, ecological and plant health conditions for which they are intended. The diversity of these conditions in the Community leads to several thousands of different products being authorised in the Member States.

This differentiation is probably best illustrated by the fact that of the 680 active substances currently authorised at national level, only 10% find uses in all Member States and as many as 25% are authorised in only one Member State.

From these considerations, it becomes clear that in a harmonised Community system, an appropriate balance must be found between

- the need for central judgment and control to ensure the proper functioning of the internal market; and
- the need for local decisions responding to local needs and situations;

while at the same time central judgment and local decisions both must ensure a high level of protection of human, animal and environmental safety.

The envisaged Community regime

These considerations were the basis of the Commission's proposal made in February 1989.

The basic principles of the proposed regime are the following:

1. The establishment of a Community positive list of active substances, whose use in formulations may be considered a priori safe for human and animal health and to the environment. This list will be established on the basis of a single application, a single dossier, the application of agreed Community criteria and a single decision by the Commission.
2. A system of registration by Member States of individual formulations containing active substances already on the Community positive list according to common data requirements, common procedural rules and common guidelines for data evaluation. These national judgments would normally be based on efficacy and crop specific residue criteria but, exceptionally, on other safety and environmental considerations as well, if particular local conditions exist rendering use of the formulation dangerous in certain geographical regions.
3. Mutual recognition by Member States of registrations granted by other Member States to the extent that the

agricultural, plant health and environmental conditions relevant to the use of the product are comparable in the regions concerned. Thus, an applicant would be able to invoke this principle of mutual recognition, where he considers the test of comparability to be satisfied. Conflicts would be resolved either between the parties immediately concerned or, if not, by a Community procedure and decision.

Other features of the proposal are:

- a systematic link between the inclusion of an active substance in the Community list and the determination of Community maximum residue levels for the agricultural products concerned;
- a 10–year programme for re-evaluating the active substances currently on the market with a view to their progressive listing on the positive list;
- harmonised labelling and packaging provisions for plant protection products, in line with other already existing Community provisions;
- improved information exchange between Member State registration authorities to facilitate cooperation and mutual recognition;
- a system of "provisional" authorisation of formulations of new active substances by Member States, pending the Community decision on listing in the positive list;
- common rules concerning the use of plant protection products and active substances in research and development activities;
- regulation of use according to principles of integrated pest control, where possible.

Finally, I would like to underline that this new legislation should not be considered on its own, but rather as a link in a whole framework of existing or proposed Community legislations, each of which addresses pesticides at particular stages and from particular angles. This is clarified in the overview in Annex I, which indicates relevant legislations as well as, for the proposal under discussion (COM(89) 34), the references made therein to other legislations for certain specific technical questions (classification, labelling, packaging, GLP).

State of discussion

The Commission adopted the proposal in February 1989. Since then, the Council, while maintaining the essentials of the proposal, has made certain elements more precise.

In February 1991, the Parliament rendered its opinion on this proposal and accepted a number of interesting amendments. These amendments are now introduced in a new proposal from the Commission for further and final consideration in the Council, so that adoption could be expected to follow in the middle of 1991.

Conclusions

The Commission believes, having regard to the particularities of the marketing and use of pesticides, that the new legislation concerning the authorisation of plant protection products represents a balanced and coherent approach to the policy objectives of the Community in relation to pesticides, in particular when this directive is considered in the whole framework of Community legislations regarding pesticides.

In the first place, by setting strict requirements, it will ensure a high level of human health and environmental protection, both in the Community decisions on active substances and in the decisions taken by Member States on preparations.

Secondly, it will eliminate the present distortions of competition between farmers in the different Member States by ensuring the availability, under comparable conditions, of the same safe and effective products for the protection of crops from pests and diseases, essential in modern agriculture.

Also, in relation to the completion of the internal market by 1992, the legislation will eliminate trade barriers which would be incompatible with the basic principles of the Single Market, agreed in the Single European Act.

Discussion: Authorisation of plant-protection products

Q: First, a simple question: why "plant protection products"? Why not just say "pesticides"? Secondly, who determines what is "comparable" among Member States?

A: The proposal does not cover all pesticides: e.g., non-agricultural or agricultural pesticides for other purposes than plant protection are not covered by the proposal.

When an application is made, the applicant must document why there is comparability of the relevant conditions in the regions of envisaged use in the Member State where authorisation is sought and in the Member State where the authorisation has already been granted. If the Member State does not accept comparability, it shall refer the case to the Commission, indicating the reasons why comparability cannot be accepted.

Q: Unlike other speakers, this one seems to be protecting cultural interests rather than man and the environment.

A: The proposal reflects a common philosophy between DG VI and DG XI. One of the major objectives of this legislation is, as I have stated, the protection of man and the environment.

Q: Does the Commission intend to set targets for reducing the use of pesticides, as has been done in Denmark, Norway and Sweden?

A: According to the previously mentioned communication "Environment and agriculture", this measure is a first step; others could be considered. I would indicate in this context the measures on organic farming, and envisaged measures regulating the use and distribution of plant protection products.

We do not now intend to set reduction targets. We believe that the complex of measures mentioned will in fact lead to reductions in use.

Feeding-stuff legislation and feed additives

by Jean Thibeaux

The quality of animal feed determines to a considerable extent the quality of animal produce, be it milk, eggs, meat or fish. For many years, particularly since the establishment of the Common Market, one of the main objectives of the Common Agricultural Policy has been to increase farming productivity in the Community in order to guarantee the supply of foodstuffs to the consumer.

The Community was in fact able to secure supplies in a relatively short time and then turned its attention to questions relating to the quality of foodstuffs, and particularly their wholesomeness. It will therefore come as no surprise that a major effort has been made to establish a harmonised body of Community legislation dealing with animal nutrition.

Categories of legislation

Analysis of the relevant Council directives makes clear that these comprise two regulatory categories. On the one hand, we have directives governing the marketing of feeding-stuffs. These are meant to set down general rules informing farmers of the composition of feeding-stuffs and the nature of the ingredients used, while at the same time guaranteeing a healthy, reliable, high-quality product through ingredients standards and labelling requirements.

On the other hand, there are safety directives, which aim to guarantee the wholesomeness of animal feeding-stuffs. The Community has had fully harmonised regulation covering the labelling of animal feeding-stuffs since 22 January 1990. This regulation will enter into force throughout the Community as of 22 January 1992.

In the area of feed safety, the Community has taken all the measures necessary to avert any danger to human or animal health.

Even before market regimes were set up – in 1960 – the Commission set up a number of working parties charged with harmonising or even drawing up legislation. Priorities in this area included additives and control methods. In 1970, the Council adopted Directive 70/524 concerning additives in feeding-stuffs and Directive 70/373 on the introduction of Community methods of sampling and analysis for the official control of feeding-stuffs. The Directive on additives is one of the most important measures for regulating animal nutrition because it has an impact on trade in both feeding-stuffs and animal products.

Substances affected by legislation

All commercial feeds contain additives; it is therefore essential that Member States apply a single list of accepted substances to avoid impediments to trade. Two distinct types of substances are covered under the definition of additives. The first is technological additives – that is, substances which have an effect on the characteristics of feeding-stuffs. This group includes preservatives, anti-oxidants, gelling agents, emulsifiers, acidity regulators, and a whole series of other aids to manufacturing which are essential in the production of animal feed.

The second group comprises zootechnical additives – that is, substances that enhance the growth of the animals. These include oligo-elements, vitamins, provitamins and other substances which have a similar effect; growth agents, commonly designated growth promoters and coccidiostats; and other medicinal substances.

This latter category of additives will in future be regulated under veterinary medicines legislation. These medical substances, administered for purely prophylactic purposes, were provisionally covered by the additives Directive precisely because there was no Community regulation on medical products.

Basic principles

The principle underlying the regulation of additives is that only those additives listed in the annexes of Directive 70/524 may be administered to animals in feed, in compliance with the conditions set down concerning species, dosage and other requirements. Administration in any other form, such as by injection or in drinking water, is forbidden. The aim of this restriction is to facilitate control of these substances.

The sale of all zootechnical additives is strictly limited to manufacturers of officially recognised feeding-stuffs. In principle, breeders may only obtain these substances in the form of feed; this virtually excludes exceeding the correct dosage.

Authorisation procedure

Authorisation of any new additive, or the new use of a previously authorised additive, must comply with five basic conditions. First, the additive must be efficient – that is, it must have a favourable effect on the characteristics of feeding-stuffs or on animal produce. Secondly, the additive should not have any harmful effect on the health of the animal, on human health, on the environment or on the quality of food. In addition, the additive must be able to be controlled; it may not be for medical or veterinary use only; and it may not have a therapeutic or prophylactic effect. The last condition is at present not applicable to coccidiostats.

These five conditions make it clear that safety of use is the dominant preoccupation of the relevant legislation: four conditions of the five refer to health aspects.

All these requirements must be verified by laboratory or field studies, the requirements for which are very specifically defined in Council Directive 87/153, which fixes the guidelines for assessing additives in animal nutrition.

Without going into detail, I would note the importance of such tests in guaranteeing the safe use of an additive. The tests required for target species include:

- toxicological studies comprising tolerance tests studying biological, toxicological, macroscopic and histological effects to determine the safety factor;

- microbiological studies, including investigation of the microbiological spectrum of action of the additive, its cross-resistance to therapeutic antibiotics, its ability to select resistance factors, its effect on intestinal flora and on the colonization of the intestinal tract, and the percentage of bacteria resistant to the additive;
- studies of the metabolism and residues of the active substance, aimed at evaluating metabolism and metabolic balance and including analytical and pharmacokinetic studies as well as studies of the bioavailability of residues.

The study on excreted residues is another important element of the dossier. A great variety of information is required in order to assess the nature and concentration of residues; persistence and kinetics of elimination; effects on methanogenosis; degradation; their effects on soil fauna, terrestrial plants, aquatic life and flora; and their toxicity in non-vertebrates and fish.

Required laboratory studies on animals include acute toxicity, mutagenicity, sub-chronic toxicity, chronic toxicity and reproductive toxicity.

Nor is this list of tests exhaustive. Depending on the nature of the active substance, further studies may be required.

All the data resulting from these studies are assembled in a dossier which must officially be submitted by a Member State to all the other Member States and to the Commission to be considered by the committee of additive experts set up by the Commission to examine authorisation requests for additives, as well as the Standing Committee for Feedingstuffs, which was set up by the Council of Ministers.

Currently, authorisation of an additive throughout the Community takes approximately four years to process. This may seem to be rather long, but it is actually very little time considering the importance of an authorisation valid in all 12 Member States and which will give rise to a number of further authorisations in third countries which align themselves with Community legislation in the field of additives.

Before submitting a dossier to the Standing Committee, the Commission generally consults the Scientific Committee on Feedingstuffs. As opposed to the Additives Committee, which is made up of government experts, the Scientific Committee is composed of scientists chosen for their expertise and, above all, independent of business and Member governments. In fact, the Scientific Committee plays a very important role in the authorisation process precisely because of the impartiality of its opinions. There is a considerable convergence of views between the government experts and the Scientific Committee.

Decisions to approve are adopted via directive by the Commission when the Standing Committee has given a favourable opinion on a request for authorisation. Opinion is by qualified majority of the Member States.

If the vote is not in favour, the Commission refers the matter to the Council, which must take a decision within three months. After that period of time, the decision may go back to the Commission unless the Council has rejected the Commission's proposals by simple majority.

An effort has been made to set up a highly efficient system, a confidence-inspiring system, organised so as to ensure that a decision will almost always be obtained.

Since its adoption in 1970, the additive Directive has been amended 69 times, for what we call adaptations to technological progress. These mostly concern allowable dose.

Permits

It is important to note that the Directive allows for two types of permit. National permits are given on a provisional basis for a maximum of five years, so that the efficacy of the product may be verified in the field. The second permit is a Community permit given to those additives which comply with all the requisite conditions of the Directive. Proof that the product will not harm humans, animals or the environment is required for both types of authorisation.

Safeguard clause

There is a safeguard clause which allows a Member State to forbid an additive or to modify its conditions for use if it judges that a condition that led to approval is no longer being fulfilled.

Such decisions are rare; the safeguard clause has been applied only once since 1970, for a preserving agent used in cat food. The procedure has never been used for additives in commercial use.

There are precise rules governing recourse to this procedure, which allow the Commission to take a decision very quickly after renewed consultation with the Standing Committee on Animal Nutrition to assess the validity of measures taken at the national level.

Labelling

Additives and premixtures of additives, as well as animal feeding-stuffs containing additives, are all subject to very strict labelling rules in order to avoid any inappropriate use of the additive and precisely to inform breeders that the feeding-stuff being sold them has certain characteristics which indicate that it must be used under very specific conditions and for a precise purpose.

Controls

In compliance with Directive 70/373, Member States have recourse to a number of analytical methods which can be used to identify additives in feeding-stuffs and to check that the authorised dose is being respected, as well as precise lists of authorised additives. These methods and lists may be applied by the Commission at the Community level, following consultation with the Standing Committee on Animal Nutrition.

The Directive also establishes a monitoring system for the marketing of additives from the time of manufacture to inclusion in feeding-stuffs.

This system includes approval by official bodies of manufacturers of additives, premixtures and feeding-stuffs. Approval depends on respect for the conditions applying to qualified personnel, equipment and installations set down in the Directive.

At the production level, the quantities of additives, premixtures and supplements must be specified, as well as the quantities delivered to users.

To ensure that products are delivered only to approved producers, Member States must annually publish a national list of manufacturers. Third-country operators are covered by this system as well.

Undesirable substances and products

The wholesomeness of animal feed-stuffs depends largely on the absence, or very low-level presence, of contaminants. These substances or products are considered undesirable because of their toxicity for the animals consuming them as well as the consumers who subsequently ingest them in the form of meat and milk.

These substances, of widely diverse origin, occur naturally in the ingredients found in feeding-stuffs and are therefore almost impossible to eliminate completely.

These substances include heavy metals, such as lead, mercury, and cadmium; mycotoxins such as aflatoxin; alkaloids and glucoids, found in some plants; and residues of particularly persistent pesticides such as organochlorides.

Directive 74/63 aims to restrict the presence of these contaminants to a level acceptable in terms of human and animal health.

Initially, Community legislation set out merely to set thresholds for each contaminant according to type of feeding-stuff and animal species. This was not a simple exercise, given the diversity of the species involved: such legislation related not merely to live-stock and poultry, but also to dogs and cats, among others. Moreover, significant metabolic differences among species made individual adjustment of limits necessary.

In recent years, a strong tendency has been noted in certain Community countries towards farm use of raw materials which are in some cases strongly contaminated by toxic substances. In 1986, this new situation prompted a widening of the scope of the Directive. As a result, Community legislation will in future cover raw materials used in the preparation of animal feeding-stuffs as well as the feeding-stuffs themselves.

In order to avoid unsuitable use by farms, the Directive states that, as of 1989, when a raw material reaches a given threshold level of an undesirable product or substance, it may not be sold to breeders in the form of a raw material. The raw material, whose toxic substances content will also be limited, as in the case of cadmium or aflatoxin, may only be made available to the animal feeding-stuffs industry, which is better equipped to carry out the necessary dilutions. Towards this end, Community regulations require specific labelling.

Import controls

In order to avoid unscrupulous importers being able to import non-compliant goods into the Community by using multiple entry points, Directive 74/63 sets up a rapid information system which simultaneously informs the inspection services of all Member States of all cases of rejections, at the same time providing all pertinent information to the departments concerned.

This rapid information procedure has been used on several occasions and does in fact dissuade. The result is that when a cargo is rejected in Hamburg, the importer will not, as was previously the case, try to have it cleared through Amsterdam or Antwerp.

Strengthening existing legislation

In the light of experience, and in particular following a regrettable accident which occurred in 1990 – although without posing a threat to public health – the Commission has decided to strengthen the existing system. In fact, it became clear that in some cases, Member State authorities had not been aware of rejection decisions and, more important, decisions to destroy batches of feeding-stuffs.

When a company charged with destroying batches does not carry out its contract, raw material that is highly dangerous to animal and human health may find its way back into the feed circuit in the form of raw materials. To avoid such a situation, the Commission, at the request of the European Parliament, is preparing a draft amendment to the Directive.

This should substantially reinforce the current information system by including all operators in the economic sector. The draft amendment will also widen the scope of the Directive to include feeding-stuffs given to animals living in nature.

At first glance, this may seem relatively unimportant, but the competent authorities believe that the habit of distributing feed to birds, animals in nature reserves and game in general is becoming more widespread and that the feeding-stuffs in question must therefore meet the health requirements laid down for domestic species.

Safeguard clause

As in the case of additives, Directive 74/63 contains a safeguard clause. A Member State may apply this clause if it believes that a non-Annex substance or product, or an existing threshold value, poses a danger to animal or human health or the environment, on the basis of new data or a reassessment of existing data.

In this case, the Member State may provisionally reduce the threshold for the substance, or even ban it, pending a Community decision.

The Directive provides that Member States must regularly carry out random checks to ensure compliance. These controls are carried out by the authorities in the Member States.

Conclusions

The Directives introduced to regulate the presence and use of certain chemical substances in foodstuffs have generally been successful - that is, to ensure that products from animals raised on feeding-stuffs manufactured by the European industry are healthy. Official surveys have shown that Community regulations are on the whole correctly implemented. In addition, an ethic appears to be developing at all levels of the process which gives reason to believe that imperfections remaining in the system will soon disappear.

Unfortunately, this does not completely preclude fraudulent acts, but these are often carried out at the level of parallel feed production circuits which are not part of the industry.

An important task that remains to be done is the drawing up and realisation of control methods. One of the first acts of Council in the area of animal nutrition was to establish the principle that feeding-stuffs should be officially controlled according to Community methods of sampling and analysis. A dozen directives, establishing over 60 analytical methods for monitoring additives, undesirable substances and ingredients have since been enacted.

Obviously, these directives do not cover all the products that are subject to Community norms. At present, the Commission is experimenting with a new working method aimed at more rapidly finalising control methods. In addition, a new directive is planned which would specify the procedures for official controls by Member States, including:

- the location of controls: these should be carried out at destination, or nearby, whenever possible;
- the nature of controls: annual or pluri-annual plans, as needed, should be drawn up;
- frequency of controls.

The Council will also certainly have to regulate the use of new substances as they are developed, as well as of products that may generally have been considered innocuous but which, in the light of new studies, are shown to be dangerous.

An example of this is provided by ensilage agents, for which regulations are currently being drawn up. The use of micro-organisms and bacteria to ensilage enzymes, or other fodder, is now governed by strict rules forbidding simple chemical substances such as sulphuric or chlorhydric acid.

Thus, while it is clear that Community regulation in this area may still be improved, it has attained its objectives of guaranteeing safety.

Discussion: Feeding stuff and feed additives

Q: What is being done about copper concentrations to prevent long-term contamination?

A: The threshold limit of 250 mg has been reduced to 150 mg after phytotoxicity was observed in the Netherlands on land with a high density of pigs.

Q: Would it not be safer for the consumer if sick animals were treated in lieu of automatically adding medicines to feed?

A: It's difficult to say. This is a question for the Council, and the Commission is planning a communication to the Council on the subject before April. According to studies, the use of these chemicals has no medical effect. In fact, their use to boost productivity is clearly distinct from medical use.

The Community policy on cosmetics

by Dimitri Angelis

The Consumers Policy Service is responsible for managing Community provisions in the cosmetics sector. These provisions are contained in a 1976 basic Directive which has been amended five times and which was adapted to technical progress on 13 different occasions. A proposal for a 6th amendment to the Directive was adopted by the Commission back in February and is at present being discussed by Community institutions.

The basic Directive is along two lines – health and consumer information. Consumer health is protected by checks on some 350–odd substances, colourings, preservatives and flavours. There are an additional 400 substances or thereabouts which are on a negative list. This is a list of substances which should not be in cosmetics.

But the 350 listed substances, as covered by the directive, represent only a very small number of those substances regularly used in cosmetics. There are in fact some 8000 substances, both of chemical and natural origin, used in cosmetics, although we cannot actually provide any statistics on the way in which those 8000 are shared out between the natural and the chemical side – at this stage anyway. Thus, 350 out of 8000 substances have been regulated.

Labelling

There are also labelling conditions. Five points are compulsory on labels for cosmetics: the name and address of the manufacturer, nominal quantity, the expiry date, the batch number and precautions on use. The label does not have to list the ingredients used. There are a certain number of provisions included in the text as well: for example, the safeguard clause – something one regularly comes across in text dealing with health matters. There are also provisions on the adaptation of the annexes – in other words, adaptation to scientific and technological progress in relation to these 350 substances.

Principles

Basically, the Directive was built on a very straightforward foundation, with which we have lived since 1976. Twelve years on, we started to rethink and to try to assess how, or just to what degree, the system was working correctly. In other words, did it provide enough protection of the citizens' health and did it, at the same time, provide the consumer with enough information?

We also tried to see to just what degree the Directive was correctly implemented in Member States. We carried out a very detailed survey on implementation of this Directive, and it proved that four Member States, at least, had adopted measures which were actually more stringent than the provisions established by the Directive – France, Spain, Portugal and Greece. These four countries have set up preliminary procedures for marketing of cosmetics.

That went against the Directive because the only thing which had to be respected by a manufacturer was the list of substances in the Directive, if the product was going to be placed on the market. Thus, every time a producer used a preserving agent, a colouring or an ultra-violet filter, he had to draw the substance he used from those listed in the Directive. In this way, he could not use any other preserving agent, UV filter or colouring. He had to select among those already listed in the Annex to the Directive; nor was he allowed to use substances from Annex II. Those were the only two provisions which are binding upon a cosmetics manufacturer in the Community.

Otherwise, manufacturers can use any other substance, at their own responsibility of course, and market their product without going through any sort of authorisation procedure imposed by the public authorities. We established that there were problems in checking implementation of this Directive. Certain legal steps were being taken, and there were certain legal points to be respected as well under Community procedures.

Implementation

In the light of all this, we tried to explain why this Directive was not being adequately implemented. We saw that certain Member States had actually exceeded the Directive's provisions, and we established that they had done so because the provisions in the Directive did not adequately protect citizens' health and did not adequately inform the consumer. Consequently, we drew up a draft directive – the 6th amendment.

In order to correct these control problems, we feel that we will have to render the cosmetics market more transparent. This means that the consumer, before he buys a product, must be aware of the ingredients in that product, which in turn means that the label must list the ingredients used in the product. Similarly, we would have to draw up a list of cosmetics ingredients: in this way, we would have a very clear overall picture of the over 8000 ingredients used.

Nomenclature

In order to facilitate intra-Community trade and limit industry costs, because obviously industry would probably have to translate the packaging labels into nine languages every time they exported, we will also have to draw up common nomenclature for these ingredients. There would then be no need to translate into the nine Community languages. We will thus require greater transparency both for the consumer and for industry.

Animal testing

Since 1979, we have been faced with requests from various associations which feel that we should either completely rule out animal testing for cosmetics or stop animal testing on so-called "decorative" cosmetics. What we have established under the 6th amendment did not satisfy either of those two requests from these associations. At this stage, we feel it is a bit unrealistic to impose an out-and-out ban on animal testing for cosmetics, either for all cosmetics or for these so-called "decorative" cosmetics.

We believe that the only way in which we could react to requests to rule out or limit animal testing would be to develop alternative methods. Until we have actually been able to do that, and until we have tried and tested methods, we feel it would be irresponsible to allow the marketing of products that we are not convinced are 100% safe.

Discussion: Cosmetics policy

Q: If you foresee a common nomenclature to avoid the need for translation, how will you ensure that consumers will be able to read labels?

A: Consumer information of this type is a special beast. A consumer will not experience side effects or allergies until after having used a product, at which point he will remember the name and/or ingredients of that product and avoid them in future.

Q: You have given excuses as to why it is difficult to provide chemical information to consumers and workers. Is it not possible to undertake a policy to make chemicals generally more transparent, with everyone their own doctor?

A: This is the goal of the 6th amendment. See Article 2.

Q: For certain cosmetics, could professionals not predict the effects on customers? What about allergies that could appear years after contact?

A: There is no horizontal approach to allergies. We are trying to handle this problem via labelling. If it becomes clear that there are good reasons for it, global regulation may in fact become necessary.

Q: If one agrees with you that animal testing cannot be entirely abolished, is it not still the case that a number of substitutes have been validated and published? Why can directives not include guidance on such screening tests?

A (Mr. Murphy): The Commission will shortly be establishing a centre for alternative testing methods with the objective of validating these for introduction into EC legislation.

My personal experience has been that there is indeed an alternative for every test. These must, however, be repeatable in every European laboratory, as well as internationally.

We have tried an alternative for LD_{50}; it cost 100 000 ECU. Thirty-four labs tested it, but it still has not been accepted by the OECD because Japan and the US are not happy with it. The result of this is that European manufacturers will have to test on animals for export purposes. Thus, we need as wide as possible acceptance for alternatives; otherwise, the result will be increased animal testing.

Annex I

List of Community Legislation

Directive 67/548/EEC on the approximation of the laws, regulations and administrative provisions relating to the classification, packaging and labelling of dangerous substances:

- 6th amendment: Council Directive 79/831/EEC
- 7th amendment Commission Proposal COM(89) 575.

Directive 74/63/EEC on the fixing of maximum permitted levels for undesirable substances and products in feeding stuffs.

Directive 75/440/EEC concerning the quality required for surface water intended for the abstraction of drinking water in the Member States.

Directive 76/895/EEC relating to the fixing of maximum levels for pesticide residues in and on fruit and vegetables.

Directive 78/631/EEC on the approximation of laws of the Member States relating to the classification, packaging and labelling of dangerous preparations (pesticides).

Directive 79/117/EEC prohibiting the placing on the market and use of plant protection products containing certain active substances.

Directive 80/68/EEC on the protection of groundwater against pollution caused by certain dangerous substances.

Directive 80/778/EEC relating to the quality of water intended for human consumption.

Regulation (EEC)797/85 on improving the efficiency of agricultural structures.

- Title V: Aids in environmentally sensitive areas from an ecological and landscape point of view.

Directive 86/362/EEC on the fixing of maximum levels for pesticide residues in and on cereals.

Directive 86/363/EEC on the fixing of maximum levels of pesticide residues in and on foodstuffs of animal origin.

Directive 86/609/EEC on the approximation of laws, regulations and administrative provisions of the Member States regarding the protection of animals used for experimental and other scientific purposes.

Directive 87/18/EEC on the harmonisation of laws, regulations and administrative provisions relating to the application of the principles of good laboratory practice and the verification of their applications for tests on chemical substances.

Regulation (EEC)1734/88 concerning export from and import into the Community of certain dangerous chemicals.

Directive 88/320/EEC on the inspection and verification of Good Laboratory Practice (GLP).

Directive 90/642/EEC on the fixing of maximum levels for pesticide residues in and on certain products of plant origin, including fruit and vegetables.

COM(88) 484 – Proposal for a Council Directive on the freedom of access to information on the environment.

COM(89) 34 – Amended proposal for a Council Directive concerning the placing of EEC-accepted plant protection products on the market.

COM(89) 552 – Proposal for a Council Regulation on organic production of agricultural products and indications referring thereto on agricultural products and foodstuffs.

COM(90) 227 – Proposal for a Council Regulation on the evaluation and the control of the environmental risks of existing substances.

COM(90) 366 – Proposal for a Council Regulation on the introduction and the maintenance of agricultural products methods compatible with the requirements of the protection of the environment and the maintenance of the countryside.